NATURE-BASED SOLUTIONS

FOR HEALTHCARE

Addressing
Anthropocene
Climate Change

DALE J. BLOCK

ISBN: 979-8-89633-053-0 (sc)
ISBN: 979-8-89633-054-7 (e)

PAGE
SOLUTIONS
Page Solutions
541 Buttermilk Pike
Crescent Springs, KY 41017

Printed in the United States of America

CONTENTS

With gratitude...

Nature-based Solutions in Healthcare: Addressing Anthropocene Climate Change afforded me the unique privilege of collaborating with my son, Jeremy, whose expertise as an environmental scientist and eco-restoration specialist brought unparalleled depth and perspective to this work. Jeremy's insights not only enriched the scientific rigor of this book but also illuminated the intricate nuances of nature-based solutions, underscoring their indispensable role in mitigating and adapting to the existential challenges posed by man-made climate change. His guidance empowered me to explore the subject with greater precision and ambition, enabling me to present readers with a comprehensive and thought-provoking examination of this critical pathway.

Writing this book alongside Jeremy has been an incredibly rewarding experience, blending intellectual discovery with personal growth. His patience and ability to articulate complex ecological and environmental concepts fostered my deeper understanding of nature's role in restoring balance to our ecosystems and healthcare systems alike. Together, we aimed to craft a narrative that is not only informative but engaging—a call to action that captures the urgency of integrating nature-based approaches into our collective response to the Anthropocene epoch.

Thank you, Jeremy, for your unwavering love and support, invaluable expertise, and shared commitment to making this book an insightful and exciting contribution to the global discourse on sustainability and resilience.

PREFACE

The convergence of healthcare and Anthropocene climate change presents one of the most urgent challenges of today. Rising global temperatures, escalating extreme weather events, and accelerating environmental degradation are placing individuals and communities at greater health risks. The consequences are far-reaching, from the immediate effects of heatwaves and air pollution to broader challenges such as shifting disease patterns, food insecurity, and growing mental health burdens. As both a frontline responder and a key stakeholder in mitigation and adaptation, the healthcare sector must actively engage in addressing these Anthropocene evolving threats.

Nature-based solutions (NbS) offer a promising and often underutilized pathway for addressing these challenges. By harnessing the power of ecosystems' services—forests, wetlands, urban green spaces, and regenerative agricultural practices—NbS can provide cost-effective, sustainable, and resilient strategies to improve health outcomes while reducing the healthcare sector's environmental footprint. The potential benefits extend beyond individual and community health, wellness, and well-being to include strengthened healthcare systems' infrastructure, reduced operational costs, and increased population health resilience.

Nature-based Solutions for Healthcare: Addressing Anthropocene Climate Change explores the role of nature-based solutions in global healthcare services delivery through a multidisciplinary lens, drawing upon evidence from environmental science, public health, healthcare policy, and clinical medicine. It examines how integrating NbS into global healthcare delivery systems can mitigate climate change impacts, support ecosystem services, and enhance human health. Case studies from around the world illustrate practical applications, from hospital gardens that promote patient recovery to large-scale green infrastructure projects that mitigate urban heat islands.

Healthcare professionals—including policymakers, urban planners, healthcare providers, and environmental advocates—serve as key stakeholders in integrating nature-based solutions into person-centered healthcare practices and public health policies. This book seeks to bridge the gap between theory and practice by offering actionable insights for implementing sustainable healthcare strategies. As we confront the dual challenges of climate change and public health protection, the adoption of nature-based solutions presents a transformational opportunity to build a more resilient, equitable, and sustainable healthcare delivery system worldwide.

I invite you to read *Nature-Based Solutions for Healthcare: Addressing Anthropocene Climate Change* with an open mind and a willingness to become part of the solution to mitigating and adapting to Anthropocene climate change. The future of integrated, person-centered healthcare systems worldwide depends not only on technological and medical advancements but also on our collective ability to embed sustainable, nature-based solutions into healthcare delivery systems at local, national, and global levels. By adopting NbS that restore ecosystems, enhance public health, and strengthen climate resilience, policymakers, clinicians, administrators, and community leaders can transform the healthcare sector into a model of environmental stewardship. Now more than ever, it is essential to recognize that human health is inseparable from the health of our planet. Through collaboration, innovation, and shared responsibility for Earth's natural resources, we can build healthcare delivery systems that are not only effective but also sustainable for generations to come.

Dale J. Block MD, MBA

May, 2025

1.0

Introduction.

1.1 Importance of Addressing Anthropocene Climate Change in the Healthcare Sector.

The healthcare sector stands at a critical juncture where environmental responsibility, person-centered high-quality care, and population health management must converge to address one of the most significant determinants of health outcomes: Anthropocene Climate Change. Empirical evidence has firmly established that climate change is no longer a distant threat but an immediate and existential crisis with profound implications for global healthcare systems and the health of all living beings. As stewards of both conservation medicine and public health, healthcare providers and stakeholders must confront the acute and chronic medical conditions exacerbated by rising temperatures, extreme weather events, and environmental degradation. These climate-driven factors are not only intensifying disease burdens but also deepening health inequities and straining an already overextended global healthcare infrastructure (Watts et al., 2018). To ensure that healthcare delivery remains resilient, equitable, and capable of protecting both present and future generations, urgent and coordinated action is imperative.

As the custodians of healthcare service delivery, medical professionals and healthcare leaders worldwide must recognize that addressing today's environmental and public health challenges requires solutions that extend beyond the confines of hospitals and clinics. Nature-Based Solutions (NbS) offer a transformational approach by integrating ecological principles with healthcare system strategies to enhance resilience, improve patient outcomes, and reduce the sector's environmental footprint (Cohen et al., 2020). By harnessing eco-systems' services—such as green infrastructure, sustainable resource management, and biodiversity conservation—NbS not only mitigates climate risks but also fosters health equity, promotes both individual and community wellness and well-being, and ensures resilience and sustainability for future generations world-wide.

To fully embed sustainability within healthcare systems, many healthcare and healthcare-related organizations world-wide have already aligned with the sustainability framework. Sustainability provides a structured methodology for healthcare institutions to assess and address their environmental stewardship, social responsibility, and ethical leadership and governance in strategic decision-making (Jameton & Pierce, 2021). By integrating sustainability principles into the core infrastructure and daily operations of local healthcare delivery systems, organizations can advance decarbonization, the process of reducing greenhouse gas (GHG) emissions through built and natural infrastructure changes, responsible supply chain management, and equitable access to high-quality healthcare service delivery.

Nature-Based Solutions (NbS) for Healthcare: Addressing Anthropocene Climate Change examines how sustainability frameworks, environmental stewardship, social responsibility, and ethical leadership and governance can serve as a strategic blueprint for healthcare institutions, hospitals, clinics, and policymakers striving to meet globally recognized sustainability performance indicators. By adopting sustainability-driven policies and integrating NbS, healthcare systems worldwide can demonstrate that environmental stewardship and high-quality, person-centered care are not only compatible but essential to the future of accessible and affordable global health for all.

Transforming healthcare world-wide requires innovative, integrated healthcare systems that address both clinical outcomes and environmental sustainability. Integrated Systems of Health (ISH) and Primary Health Care (PHC) provide essential frameworks for advancing nature-based solutions in the global pursuit of high-quality, person-centered healthcare services delivery. ISH emphasizes seamless coordination across healthcare networks, embedding environmental sustainability into healthcare models to improve individual and community health outcomes (Horton et al., 2022). PHC, encompassing advanced primary care services, essential public health functions, and community-based interventions, prioritizes proactive health measures such as disease prevention, addressing health-related social needs, and promoting health equity. By incorporating NbS—such as urban green spaces, sustainable facility design, and nature-based therapeutic interventions—PHC can help reduce disease burdens and enhance mental well-being (World Health Organization [WHO], 2021).

By embedding NbS within ISH and PHC, healthcare delivery systems can shift from reactive, high-cost treatment models of acute and chronic condition care to sustainable, community-oriented primary health care strategies that emphasize proactive models of health (i.e., health protection, prevention, promotion, and preparedness), producing positive health outcomes with optimized wellness and enhanced well-being. *Nature-Based Solutions (NbS) for Healthcare: Addressing Anthropocene Climate Change* highlights case studies demonstrating the global effectiveness of these approaches, illustrating how integrating ecosystems' services into healthcare service delivery and healthcare delivery systems can foster resilience in marginalized and vulnerable populations while mitigating and adapting to Anthropocene climate change.

Beyond the delivery of essential healthcare services, climate resilience must also extend to the workplace, ensuring that both healthcare and non-healthcare facilities provide safe, sustainable environments for employees and employers alike. Environmental,

Health, and Safety (EHS) programs play a vital role in embedding nature-based solutions (NbS) within these institutions, addressing workplace hazards, air and water quality, and resource management (Patel et al., 2023). Sustainable building designs, energy-efficient operations, and waste reduction strategies are key ways EHS programs can integrate NbS to create healthier work environments while minimizing ecological impact.

By incorporating NbS into EHS initiatives, leaders across healthcare and related sectors, along with organizational governance, can mitigate climate-related risks such as heat stress and air pollution while fostering a culture of sustainability. This book explores the practical applications of NbS in global workplace safety programs, illustrating how investments in sustainable infrastructure and green technologies contribute to long-term health and economic benefits.

The transition to world-wide, sustainable healthcare services' delivery is not just a moral and ethical imperative—it is a strategic necessity to ensure long-term health security and healthcare delivery system resilience. Anthropocene climate-change effected stakeholders around-the-world including professionals, policymakers, and innovators must act now! Through a multidisciplinary lens, Nature-based Solutions (NbS) provide actionable strategies to bridge the gap between climate science and traditional clinical medicine and public health practices.

1.2 Why Healthcare Must Act Now: Beyond a Burning Platform.

The urgency for the global medical-industrial complex to address Anthropocene climate change extends far beyond a reactionary response to a crisis—it is a strategic imperative for long-term healthcare system sustainability and resilience. The industry's role in mitigating climate change is particularly significant given its growing environmental footprint. Globally, the healthcare sector is responsible for 4.4 % of total global GHG emissions, a figure that continues to rise due to high energy consumption, medical waste production, and reliance on resource-intensive supply chains (Karliner et al., 2019). From energy-intensive hospital operations to the disposal of pharmaceuticals and medical plastics, the global medical-industrial complex's impact on environmental degradation is undeniable.

At the core of this challenge lies a profound ethical responsibility. Healthcare delivery systems are designed to provide proactive services, manage acute and chronic conditions, and address upstream health-related social needs. However, current operational and management practices contribute to the very environmental crises that worsen these conditions. Pollution-related health issues—such as respiratory illnesses, heat-related morbidity, and waterborne diseases—disproportionately impact marginalized and vulnerable populations, many of whom already face inequities in healthcare access (Ebi et al., 2021). If the medical-industrial complex, along with global healthcare leaders, fails to collaborate, coordinate, cooperate, and communicate (the 4 C's) and take immediate action, they risk perpetuating cycles of environmental harm that undermine clinical medicine, public health, and the global pursuit of improved health, wellness, well-being, and resilience.

The inextricable interconnectedness of planetary health, human health, and environmental health protection cannot be overstated. The degradation of ecosystems directly impacts human health through increased vector-borne diseases, food insecurity, and extreme weather-related health emergencies (Whitmee et al., 2015). Conversely, investing in NbS—such as green hospital design, sustainable medical supply chains, and ecological restoration projects for climate resilience—can drive positive health outcomes while simultaneously safeguarding the environment and local ecosystems. Addressing Anthropocene climate change is not a peripheral issue for healthcare; it is central to the sector's ethical and operational responsibilities.

This book argues that healthcare must move beyond a "burning platform" short-term approach—where action is taken only in response to immediate crises—and instead embrace a long-term, proactive, sustainability-driven transformational change. By integrating sustainability frameworks, leveraging Nature-based Solutions (NbS) within Integrated Systems of Health (ISH) and Primary Health Care (PHC), and embedding climate-conscious strategies in Environmental, Health, and Safety (ESH) programs, healthcare can lead the way in building a resilient, sustainable future for everyone on the planet Earth.

1.3 Scope of the Book.

Nature-Based Solutions for Healthcare: Addressing Anthropocene Climate Change examines the essential role healthcare delivery systems play in confronting the global existential threat of Anthropocene climate change by incorporating nature-based solutions (NbS) into medical practice, healthcare infrastructure, and organizational governance. This book offers a comprehensive framework for utilizing ecosystems' services to create climate-resilient global healthcare systems, reduce environmental impact, and advance integrated person-centered primary health care (IPC-PHC) in response to growing climate challenges.

This book is designed to provide both theoretical insights and practical applications, catering to a diverse range of healthcare stakeholders, including healthcare professionals, policymakers, hospital and health system administrators, and sustainability advocates. It adopts a multidisciplinary approach, grounded in empirically-supported and evidence-based principles from environmental science, public health, healthcare management, health systems science, and ecological design. Through case studies, policy recommendations, and data-driven strategies, this book serves as a comprehensive guide for transforming healthcare delivery systems worldwide into leaders in climate resilience and sustainability.

Below are the key themes and areas of focus:

1. *The Intersection of Climate Change and Healthcare.*
   a. The growing impact of climate change on human health, including the rise in climate-sensitive diseases, mental health challenges, and health inequities.
   b. The medical-industrial complex's role as both a contributor to and a solution for environmental degradation.
   c. The necessity for urgent action beyond reactive crisis management to a long-term, sustainable transformation.

2. *Nature-Based Solutions (NbS) in the Global Delivery of Healthcare Services.*
   a. NbS solutions and the relevance to healthcare settings geographically (i.e., HIC and LMIC).
   b. How ecosystems' services can be integrated into hospital design, patient care, and community health initiatives.
   c. The potential of NbS to enhance healthcare delivery system resilience while reducing operational costs and ecological impact.

3. *Sustainability in Global Healthcare Delivery Systems.*
   a. The importance of sustainability principles (i.e., environmental stewardship, social responsibility, and ethical leadership and governance) in guiding healthcare organizations world-wide toward sustainable practices.
   b. Strategies for incorporating sustainability principles into healthy public policies and operations around-the-world.
   c. The financial benefits of aligning healthcare institutions with global sustainability goals.

4. *Integrating NbS into Healthcare Delivery Models: Primary Health Care (PHC) and Integrated Systems of Health (ISH).*
    a. The role of NbS in strengthening Primary Health Care (PHC) through community-based interventions, green public spaces, and sustainable healthcare facilities.
    b. Why Integrated Systems of Health (ISH) should incorporate NbS to enhance patient outcomes, proactive care, and climate resilience.
    c. Report out on case studies demonstrating successful implementation of NbS in diverse healthcare delivery settings.

5. *NbS in Environmental, Health, and Safety (ESH) Programs.*
    a. NbS in ESH Programs: Historical context and current state.
    b. Comparisons between EHS and Sustainability Programs and their alignment.
    c. Climate conscious EHS.
    d. The role of green infrastructure in mitigating heat stress, improving air quality, and reducing exposure to harmful pollutants.
    e. Sustainable resource allocation and management in hospitals, including water conservation and energy efficiency.

6. *Policy, Advocacy, and Global Leadership in Sustainable Healthcare.*
    a. The role of healthcare leaders in advocating for policies that integrate NbS into national and global health strategies.
    b. What international climate-change frameworks, such as the Paris Agreement and the UN Sustainable Development Goals (SDGs), align with global healthcare delivery systems' sustainability vision.

   c. The need for interdisciplinary and multisectoral 4 C's (collaboration, cooperation, coordination, and communication) amongst national and local governments, healthcare institutions, and environmental organizations.

1.4 A Roadmap for Action.

The final section of the book outlines a call to action, providing a roadmap for healthcare organizations, policymakers, and clinicians world-wide to adopt nature-based solutions (NbS) at scale. It highlights critical steps for integrating sustainability into medical education, patient care, research, and healthcare administration while reinforcing the sector's ethical obligation to lead in climate resilience.

By framing healthcare as a key driver of climate solutions, *Nature-Based Solutions for Healthcare: Addressing Anthropocene Climate Change* positions the sector as an essential partner in building a healthier, more sustainable world for future generations.

This book is designed for:

1. Healthcare professionals seeking to understand the impact of Anthropocene climate change on patient care and healthcare systems' operations.
2. Hospital administrators and policymakers looking for sustainable and cost-effective healthcare solutions.
3. Public health leaders interested in integrating climate resilience into both individual and population health management.
4. Researchers and academics exploring the intersection of ecosystems' services, environmental sustainability, and healthcare service delivery.

Environmental and sustainability advocates aiming to align global healthcare delivery systems with global climate initiatives.

One final thought: The time for healthcare delivery systems worldwide to embrace nature-based solutions is now! This book offers the knowledge, tools, and inspiration necessary to shift from a reactive approach to healthcare—focused on acute and chronic care—to a proactive, climate-conscious, integrated, and person-centered global systems of health with primary health care as its cornerstone of foundation. This model aims to protect both human and planetary health, ensuring a sustainable and resilient future for all living things on Earth.

1.5 References.

1. Cohen, A. J., Brauer, M., Burnett, R., et al. (2020). The impact of climate change on human health and the role of healthcare systems in adaptation. *The Lancet Planetary Health, 4*(4), e135–e144.
2. Ebi, K. L., Semenza, J. C., & Rocklöv, J. (2021). Climate change and health: Impacts, vulnerability, and adaptation. *The Lancet, 398*(10304), 2097–2106.
3. Horton, R., Lo, S., & Kleinert, S. (2022). The case for Integrated Systems of Health in addressing the climate crisis. *The Lancet, 400*(10361), 1125–1130.
4. Jameton, A., & Pierce, J. R. (2021). Healthcare sustainability and the ESG framework: The intersection of climate action and health equity. *Health Affairs, 40*(9), 1345–1352.
5. Karliner, J., Slotterback, S., Boyd, R., et al. (2019). *Health care's climate footprint: How the health sector contributes to the global climate crisis and opportunities for action.* Health Care Without Harm.
6. Patel, D. M., Green, L., & Robinson, J. (2023). Environmental, health, and safety programs in healthcare: A pathway to climate resilience. *Journal of Occupational and Environmental Medicine, 65*(2), 89–101.

7. Watts, N., Amann, M., Arnell, N., et al. (2018). The 2018 report of the Lancet Countdown on health and climate change: shaping the health of nations for centuries to come. *The Lancet, 392*(10163), 2479–2514.

8. Whitmee, S., Haines, A., Beyrer, C., et al. (2015). Safeguarding human health in the Anthropocene epoch: Report of The Rockefeller Foundation–Lancet Commission on planetary health. *The Lancet, 386*(10007), 1973–2028.

9. World Health Organization. (2021). *Primary health care and sustainable development: Strengthening community resilience in a changing climate.* WHO Press.

2.0

The Intersection of Anthropocene Climate Change and Healthcare.

2.1 The Growing Impact of Anthropocene Climate Change on Human Health.

The term *Anthropocene* refers to a proposed geological epoch characterized by the significant and lasting impact of human activities on Earth's ecosystems and geology, including climate change, biodiversity loss, and widespread environmental degradation (Crutzen et al., 2000). It is marked by human influence becoming a dominant force in shaping the planet's physical and biological processes. This concept highlights the extent to which humanity has altered the planet, often to the detriment of natural ecosystems.

Climate change—an accelerating rise in global surface temperature driven by human activity—disrupts local temperatures, humidity, wind patterns, precipitation, soil moisture, and sea levels (Wadanambi et al., 2020). It is increasingly recognized as one of the greatest existential threats to human health in the 21st century, exacerbating health disparities and placing immense strain on healthcare systems worldwide. Rising global temperatures, extreme weather events, shifting disease patterns, and environmental degradation are contributing to a surge in climate-related illnesses and mortality (Watts et al., 2018). The health consequences of climate change are far-reaching, affecting physical, mental, and social well-being across populations.

1. *Rising Temperatures and Heat-Related Illnesses.*
    Global temperatures have risen significantly over the past century, with heatwaves becoming more frequent, prolonged, and intense. Extreme heat poses a direct threat to human health, increasing the risk of heat exhaustion, heatstroke, cardiovascular stress, and kidney disease. Vulnerable populations—such as older adults, children, pregnant individuals, and those with chronic conditions—are particularly at risk (Haines & Ebi, 2019). In urban areas, *the heat island effect* further amplifies temperature extremes, disproportionately affecting low-income communities with limited access to cooling resources.

2. *Air Pollution and Respiratory Diseases.*
    Climate change contributes to deteriorating air quality through increased wildfires, higher concentrations of ground-level ozone, and the spread of allergens such as pollen. Air pollution is a major driver of respiratory diseases, including asthma, chronic obstructive pulmonary disease (COPD), and lung cancer. Fine particulate matter (PM2.5) and nitrogen oxides (N_2O)—both exacerbated by fossil fuel combustion—are linked to cardiovascular disease, stroke, and premature death (Landrigan et al., 2018).

3. *Infectious Diseases and Changing Vector Patterns.*
    Warmer temperatures and changing precipitation patterns world-wide are altering the geographic distribution of vector-borne diseases such as malaria, dengue fever, Lyme disease, and West Nile virus. Mosquitoes, ticks, and other disease-carrying vectors are expanding into new regions, increasing the risk of outbreaks in previously unaffected areas (Mora et al., 2022). Waterborne diseases, including cholera and gastrointestinal infections, are also rising due to flooding, contamination of drinking water supplies, and inadequate sanitation.

4. *Food and Water Insecurity.*

   Climate change threatens global food security by disrupting agricultural productivity through droughts, soil degradation, extreme weather events, and shifting weather conditions suitable for pests and diseases limiting crop production. Declining crop yields and reduced nutritional quality of staple foods increase malnutrition and foodborne illnesses, particularly in vulnerable populations (Myers et al., 2017). Water scarcity and contamination from pollution, rising sea levels, and industrial waste further compound health risks, leading to dehydration, kidney disease, and waterborne infections.

5. *Mental Health and Psychological Stress.*

   The psychological impacts of climate change are profound, with increasing evidence linking climate-related disasters to anxiety, depression, post-traumatic stress disorder (PTSD), and eco-anxiety (Cianconi et al., 2020). Communities affected by hurricanes, floods, tornados, wildfires, and prolonged droughts often experience displacement, economic hardship, and loss of social support networks, contributing to long-term mental health challenges. Children and adolescents are particularly vulnerable to the emotional toll of climate change, as they face an uncertain future shaped by environmental instability.

6. *Displacement, Migration, and Social Disruptions.*

   Climate change is a major driver of displacement and forced migration, as rising sea levels, desertification, and extreme chronic weather events render many regions uninhabitable. Climate refugees often face inadequate access to healthcare, food, and sanitation, leading to increased morbidity and mortality. Moreover, climate-induced displacement exacerbates social inequalities, disproportionately impacting marginalized populations and increasing the risk of conflict over limited resources (Hsiang et al., 2013).

In summary, the growing health impacts of climate change necessitate urgent action from the healthcare sector, policymakers, and global leaders. By integrating climate adaptation and mitigation strategies into healthcare systems—such as Nature-Based Solutions (NbS), sustainable infrastructure, and climate-informed public health initiatives—societies can build resilience against the escalating health threats posed by a changing climate. Addressing climate change is not just an environmental necessity but a fundamental public health imperative.

2.2 The Medical-Industrial Complex and Climate Change.

The term, *medical-industrial complex*, refers to the interconnected network of industries, institutions, and professionals that drive healthcare practices and policies, often influenced by economic interests and profit motives. This complex includes healthcare providers, pharmaceutical companies, medical device manufacturers, insurers, and other stakeholders who shape the delivery of healthcare services. While the medical-industrial complex has significantly contributed to advancements in medical technology, treatments, and care, critics argue that it has led to a system where profit generation can sometimes take precedence over patient care, equity, and public health. The prioritization of expensive, often unnecessary treatments, over prevention and integrated person-centered primary health care, is a central concern in the debate over the ethics of the medical-industrial complex.

The growth of the medical-industrial complex is tied to the commercialization of healthcare, which has resulted in a system that is deeply embedded in market-driven forces. This has led to a situation where financial incentives are often aligned with procedures and interventions that may not necessarily align with the best interests of patients. The increasing influence of pharmaceutical companies and for-profit healthcare organizations has raised concerns about conflicts of interest, with critics suggesting that these organizations exert too much power over healthcare policy and practice (Hartmann, 2016).

Additionally, the complex is often criticized for contributing to escalating healthcare costs, creating barriers to equitable access to care, and exacerbating health disparities. Reforming the medical-industrial complex requires a shift towards prioritizing patient-centered care, improving healthcare affordability, and addressing the systemic drivers of inequality in healthcare delivery (Himmelstein et al., 2016).

The medical-industrial complex plays a paradoxical role in relation to climate change. It is both a major contributor to environmental degradation and a potential driver of sustainable solutions. The industry must reconcile its mission to produce positive health, optimized wellness, and enhanced well-being with its substantial greenhouse gas (GHG) footprint and ecological impact.

The medical-industrial complex is a significant contributor to climate change in the following ways:

1. *Greenhouse Gas (GHG) Emissions and Energy Consumption.*
    The global healthcare sector accounts for approximately 4.4% of net global GHG emissions, which is comparable to the aviation industry (Karliner et al., 2019). Hospitals, which operate around the clock, consume vast amounts of energy for lighting, heating, cooling, and medical equipment. The reliance on fossil-fuel-based electricity exacerbates GHG output.

2. *Medical Waste and Pollution.*
    a. *Single-Use Plastics*: Hospitals produce vast quantities of plastic waste, including packaging, disposable gloves, syringes, and IV bags. The overuse of single-use items contributes significantly to landfill overflow and ocean pollution (Thompson et al., 2021).
    b. *Pharmaceutical Contaminants*: Medications improperly disposed of—whether through hospital waste streams or patient excretion—contaminate water sources. Antibiotics and endocrine disruptors have been detected in drinking

water, raising concerns about antibiotic resistance and eco-system damage (Daughton & Ruhoy, 2009).

   c. *Radiological and Chemical Waste*: Diagnostic imaging (e.g., MRI and CT scans) requires high energy use, while contrast agents and radioactive isotopes contribute to hazardous waste streams.

3. *Supply Chain Emissions.*
Healthcare relies on a globalized supply chain for pharmaceuticals, medical equipment, and personal protective equipment (PPE). The production, packaging, and transportation of these materials create Scope 3 emissions, which account for over 70% of healthcare's total GHG inventory (Karliner et al., 2019).

4. *Deforestation and Land Use.*
The production of pharmaceuticals often requires raw materials sourced from biodiverse ecosystems. Deforestation for medicinal plant harvesting, palm oil production (used in many healthcare products), paper and packaging production, and land conversion for healthcare infrastructure contribute to biodiversity loss and ecosystem disruption (Watts et al., 2018).

The medical-industrial complex is a significant contributor to multi-dimensional solutions to climate change in the following ways:

1. *Green Hospital Initiatives.*
   a. *Renewable Energy Adoption*: Some health systems, such as Kaiser Permanente, have committed to 100% renewable energy and net-zero strategies for GHG neutrality, setting a model for sustainability in healthcare (Kaiser Permanente, 2020).
   b. *Energy-Efficient Infrastructure*: Leadership in Energy and Environmental Design (LEED)-certified hospitals incorpo-

rate sustainable building materials, smart energy systems, and natural ventilation to reduce emissions.

   c. *Sustainable Procurement*: Implementing circular economy targets—such as 20% reduction in absolute total hazardous and non-hazardous waste generated by 2050, 50% combined recycling rate for hazardous and non-hazardous waste, and 100% zero waste to landfill by 2035—reduces waste and supply chain emissions (AbbVie, 2023).

2. *Reducing Medical Waste and Sustainable Practices.*
   a. *Waste Reduction Strategies*: Hospitals are exploring sterilization and reuse of medical equipment into a safe and reducing dependence on disposables.
   b. *Green Pharmacy Initiatives*: Strategies such as take-back programs for unused medications and green chemistry approaches in drug design help minimize pharmaceutical contamination.
   c. *Biodegradable and Recyclable Materials*: The development of compostable PPE and plant-based plastics offers alternatives to fossil-fuel-derived medical supplies.

3. *Telehealth and Digital Transformation.*
   The expansion of telemedicine during the COVID-19 pandemic demonstrated its potential to reduce healthcare-related travel emissions. By decreasing the number of in-person visits, digital health solutions can significantly cut the GHG footprint of patient care (Hollander & Carr, 2020).

4. *Climate-Resilient Healthcare Systems.*
   The climate crisis is a health crisis, with rising temperatures increasing the burden of vector-borne diseases, respiratory illnesses, and heat-related mortality (Watts et al., 2021). Climate-resilient health systems must integrate the following:

a. *Disaster Preparedness*: Hospitals must design infrastructure, both built and natural, to withstand extreme weather events.

b. *Sustainable Supply Chains*: Localized production of essential medicines and medical supplies can reduce supply chain delays due to natural disaster- induced downtime at supply hubs, and reliance on GHG-intensive global trade.

c. *Integration of Climate-Health Data*: Public health systems must track climate-influenced disease trends and adjust interventions accordingly.

The healthcare sector has a moral and strategic imperative to lead on climate action. Policies that could accelerate sustainability include:

1. *Regulatory Incentives*: Governments should provide tax breaks or funding for hospitals implementing green technologies.
2. *Emissions Reporting*: Mandatory GHG inventory and climate change risk reporting for healthcare organizations can increase transparency and drive accountability.
3. *Interdisciplinary Collaboration*: Partnerships between healthcare providers, environmental and climate scientists, sustainability consultants, and policymakers can foster cross-sectoral systemic change.

In summary, the medical-industrial complex must undergo a major paradigm shift from a major polluter into a leader of environmental sustainability. While the sector's GHG footprint is significant, proactive strategies—ranging from green hospital design to telehealth adoption—offer a pathway toward a healthier planet and healthier populations. Addressing Anthropocene climate change is not just an environmental necessity but a public health imperative.

2.3 The Necessity of Urgent Action by the
Medical-Industrial Complex.

The Anthropocene epoch, marked by rapid and irreversible human-driven environmental change, presents an unprecedented challenge to global health, wellness and well-being that the medical-industrial complex can no longer address through reactive crisis management alone. Historically, healthcare delivery systems have responded to public health threats as they emerge, such as pandemics, natural disasters, and disease outbreaks. However, climate change represents a slow-moving yet intensifying catastrophe that fundamentally alters disease patterns, infrastructure stability, and healthcare delivery (Watts et al., 2021). Extreme weather events—hurricanes, wildfires, droughts, and heatwaves—are no longer isolated occurrences but frequent and intensifying realities, straining emergency preparedness, resilience, and rapid response systems and disproportionately affecting marginalized and vulnerable populations. Hospitals and clinics, many of which are already operating near capacity, are ill-equipped to handle the cascading effects of climate-induced crises without a shift in strategy.

Beyond acute disasters, climate change is driving chronic health burdens, such as increased rates of cardiovascular disease from air pollution, higher transmission of vector-borne diseases due to loss of ecosystem functions that minimizes disease--carrying insect populations, and worsening mental health due to displacement and environmental anxiety (Watts et al., 2021). The economic cost of these health consequences is staggering, with climate-related illness and disaster recovery placing an increasing burden on national healthcare budgets. If the healthcare sector remains locked in a reactionary cycle, it will find itself perpetually overwhelmed, dealing with the symptoms of climate change rather than addressing its root causes. This is an unsustainable trajectory that demands a major paradigm shift—one in which healthcare institutions actively work to mitigate climate change impacts, enhance built, natural, and human resilience, and integrate sustainability into the foundation of integrated person-centered primary health care (IPC-PHC). The healthcare sector must recognize that environmental health and human health are inextricably linked, requiring a shift from short-term adaptation measures to long-term systemic transformational change.

For the medical-industrial complex to align its operations with its fundamental mission of healing and harm prevention, it must commit to deep, structural, and long-term sustainability initiatives rather than superficial, symbolic efforts. While many healthcare and healthcare-related organizations have made progress by adopting net-zero goals, energy-efficient buildings, and waste reduction programs, these incremental changes are small iterative changes that minimally impact or influence the sector's enormous contributions to the global environmental demise. Achieving real sustainability requires a fundamental rethinking of healthcare services delivery, medical supply chains, and integrated systems of health (ISH), focusing on reducing reliance on fossil fuels, minimizing waste, and prioritizing environmentally responsible procurement (Karliner et al., 2019). For instance, healthcare infrastructure should begin the transition to fully renewable energy sources, such as solar, nuclear, and wind power, to ensure that hospitals are reducing their GHG emissions. In addition, to remain operational during extreme weather events, moving their renewable energy sources on-site to reduce their reliance on the power generated from the grid, improves their infrastructure resilience while reducing their Scope 2 GHG emissions. Similarly, the industry must reform its dependence on single-use plastics, opting for biodegradable or reusable alternatives while maintaining the highest continuous quality and safety standards.

Greenhouse gas (GHG) emissions are categorized into Scope 1, 2, and 3 emissions based on their sources. Scope 1 emissions are direct emissions from owned or controlled sources, such as fuel combustion in company-owned vehicles or on-site energy production. Scope 2 emissions are indirect emissions from purchased electricity, steam, heating, and cooling consumed by an organization. Scope

3 emissions encompass all other indirect emissions in a company's value chain, including those from supply chain activities, business travel, and product use. Managing all three scopes is essential for organizations to achieve comprehensive carbon reduction strategies (World Resources Institute & World Business Council for Sustainable Development, 2004).

The medical-industrial complex's role in climate change cannot be confined to hospital walls; healthy public policy and upstream health-related social needs and other determinants of health reforms are essential to sustaining progress. Medical education at all levels of engagement must integrate climate-conscious healthcare practices, preparing the next generation of physicians, nurses, and administrators to consider environmental sustainability as a core component of IPC-PHC. Regulatory bodies should mandate GHG emission inventory reporting for healthcare and health-care related organizations, holding them accountable for GHG emissions and incentivizing more environmentally-friendly practices. Pharmaceutical companies and medical device manufacturers must shift toward circular economy principles, designing products that minimize waste, utilize sustainable materials, and allow for recycling or safe disposal. Furthermore, healthcare delivery systems world-wide must embrace telemedicine and digital health solutions where applicable, which not only enhance patient accessibility but also reduce emissions from transportation and resource consumption (Hollander & Carr, 2020).

Without a commitment to these long-term, systemic changes, the medical-industrial complex risks becoming one of the largest contributors to the very environmental degradation that is threatening global health. True leadership in global healthcare systems today requires acknowledging that treating the planet is as essential as treating patients, ensuring that global healthcare systems serve as a solution rather than a driver of climate change.

2.4 References.

1. AbbVie. (2023). 2023 ESG action report. AbbVie Inc.
2. Cianconi, P., Betrò, S., & Janiri, L. (2020). The impact of climate change on mental health: A systematic descriptive review. *Frontiers in Psychiatry, 11*, 74.
3. Crutzen, P. J., & Stoermer, E. F. (2000). The Anthropocene. *Global Change Newsletter*, 41, 17-18.
4. Daughton, C. G., & Ruhoy, I. S. (2009). Environmental footprint of pharmaceuticals: The significance of factors beyond direct excretion to sewers. *Environmental Toxicology and Chemistry, 28*(12), 2495-2521.
5. Haines, A., & Ebi, K. (2019). The imperative for climate action to protect health. *New England Journal of Medicine, 380*(3), 263-273.
6. Hartmann, D. (2016). The medical-industrial complex and the pursuit of profit: The role of pharmaceutical companies and healthcare providers. *Journal of Public Health Policy*, 37(4), 469-483.
7. Himmelstein, D. U., & Woolhandler, S. (2016). The medical-industrial complex in the U.S.: Historical and contemporary perspectives. *International Journal of Health Services*, 46(2), 365-378.
8. Hollander, J. E., & Carr, B. G. (2020). Virtually perfect? Telemedicine for COVID-19. *New England Journal of Medicine, 382*(18), 1679-1681.
9. Hsiang, S., Burke, M., & Miguel, E. (2013). Quantifying the influence of climate on human conflict. *Science, 341*(6151), 1235367.
10. Kaiser Permanente. (2020). Becoming greenhouse gas neutral: A model for climate-smart healthcare. *Kaiser Permanente Sustainability Report.*
11. Karliner, J., Slotterback, S., Boyd, R., Ashby, B., & Steele, K. (2019). Health care's climate footprint: How the health sector contributes to the global climate crisis and opportunities for action. *Health Care Without Harm & Arup Report.*

12. Landrigan, P. J., Fuller, R., Acosta, N. J., et al. (2018). The Lancet Commission on pollution and health. *The Lancet, 391*(10119), 462-512.

13. Mora, C., McKenzie, T., Gaw, I. M., et al. (2022). Over half of known human pathogenic diseases can be aggravated by climate change. *Nature Climate Change, 12*(9), 869-875.

14. Myers, S. S., Smith, M. R., Guth, S., et al. (2017). Climate change and global food systems: Potential impacts on food security and undernutrition. *Annual Review of Public Health, 38,* 259-277.

15. Thompson, M., Crowe, S., & Fleck, C. (2021). The environmental impact of single-use plastics in healthcare: A case for sustainable solutions. *The Lancet Planetary Health, 5*(8), e450-e455.

16. Watts, N., Amann, M., Arnell, N., et al. (2018). The 2018 report of the Lancet Countdown on health and climate change: Shaping the health of nations for centuries to come. *The Lancet, 392*(10163), 2479-2514.

17. Watts, N., Amann, M., Arnell, N., et al. (2021). The 2021 report of The Lancet Countdown on health and climate change: Code red for a healthy future. *The Lancet, 398*(10311), 1619-1662.

18. Wadanambi, R.T., Wandana, L.S., Chahumini, K.K.G.L., et al. (2020) The effects of industrialization on climate change. *Journal of Research Technology and Engineering.* 1(4), 86-94.

19. World Resources Institute & World Business Council for Sustainable Development. (2004). *The greenhouse gas protocol: A corporate accounting and reporting standard.* Washington, DC: World Resources Institute.

3.0

Nature-Based Solutions (NbS) in the Global Delivery of Healthcare Services.

As the medical-industrial complex seeks sustainable approaches to mitigate and adapt to its environmental impact, *Nature-Based Solutions (NbS)* offer a transformational strategy for integrating ecological resilience into healthcare services delivery. For the healthcare sector, NbS harness the power of ecosystems' services to address individual and community health challenges, healthcare services' barriers, improve patient outcomes, and enhance the sustainability of healthcare infrastructure. From green hospital designs to urban reforestation initiatives that reduce greenhouse gas (GHG) emissions, air pollution and heat exposure, these solutions align multi-level public health practices with environmental stewardship, creating a more resilient and adaptive healthcare delivery system.

3.1 Nature-Based Solutions (NbS) and Their Relevance in Healthcare Settings Across Geographies.

3.1.1 Introduction.

Defining Nature-Based Solutions (NbS) in healthcare refers to actions that leverage ecosystems and biodiversity conservation to address societal challenges, including Anthropocene climate change, public health crises (e.g., natural disasters and vector-borne infectious diseases), and environmental sustainability (Cohen-Shacham et al., 2016). These solutions integrate ecological processes into healthcare delivery system infrastructure and service delivery, reducing the healthcare's sector GHG footprint while improving human health outcomes.

In clinical settings, NbS can take various forms, from biophilic (i.e., biophilic design is rooted in the idea that humans have an innate affinity for nature, and that incorporating natural elements into the built environment can improve health, wellness, and well-being) hospital designs and green spaces that promote healing to wetlands that improve water security and urban forests that mitigate air pollution. These approaches recognize that human health and environmental health are deeply interconnected, and fostering ecological resilience is critical to ensuring a sustainable healthcare delivery system.

NbS are especially relevant in the face of climate change, which is already driving extreme weather events, air pollution, biodiversity loss, and the spread of vector-borne infectious diseases—all of which strain healthcare infrastructure and exacerbate health disparities (Watts et al., 2021). By embedding NbS into healthcare services planning and implementation, hospitals' operations and management, and public health systems science can enhance their resilience, reduce operational costs, and create healthier environments for both patients and healthcare workers. However, the effectiveness and implementation of NbS vary significantly between high-income countries (HICs) and low- and middle-income countries (LMICs) due to differences in financial resources, policy frameworks, and healthcare infrastructure.

3.1.2 NbS Adoption in High-Income Countries (HIC).

In high-income countries (HICs), the adoption of NbS in clinical and clinically-related settings is driven by advanced medical and informational technological innovation, healthy public policy incentives, and the medical-industrial complex's institutional investments in sustainability. With access to financial resources and research capabilities, healthcare delivery systems in HICs have the opportunity to integrate advanced NbS into hospital design, urban planning, and public health initiatives to reduce their GHG emissions, to improve climate resilience, and improve human physical and mental health. Some key applications include:

1. *Green Hospital Infrastructure.*
   Many hospitals in HICs are adopting biophilic architecture and green building principles to enhance energy efficiency, reduce waste, and improve patient recovery rates. For example:
   a. Ng Teng Fong General Hospital in Singapore is designed with extensive rooftop gardens, cross-ventilation for natural cooling, and vertical greenery that helps regulate indoor temperatures while improving air quality (Jimenez et al., 2021).
   b. Maggie's Centres in the UK, a series of cancer treatment facilities, incorporate natural light, green courtyards, and therapeutic gardens, which have been shown to reduce stress and improve mental well-being in cancer patients (Ulrich et al., 1991).
   c. The Ellen MacArthur Cancer Center in the Netherlands integrates rainwater harvesting, solar panels, and green roofs, reducing water and energy consumption while enhancing biodiversity (EMF, 2022).

   These green infrastructure strategies not only lower hospital emissions but also create healing environments that improve patient outcomes and staff wellness and well-being.

2. *Urban Reforestation for the Public's Health.*
   Air pollution is a major public health threat, contributing to respiratory diseases, cardiovascular conditions, and premature mortality. In response, many cities in HICs are investing in

urban forestry initiatives to mitigate pollution and reduce the urban heat island effect.

   a. The Green Heart Project in Louisville, Kentucky, aims to plant trees in urban neighborhoods to lower air pollution and improve cardiovascular health outcomes. Studies have shown that increased tree cover is associated with lower blood pressure and reduced stress levels (Hoffman et al., 2020).

   b. In Berlin, Germany, green corridors have been developed to filter pollutants and reduce noise pollution, benefiting both public health and biodiversity (Grün Berlin GmbH, 2025)

   c. Toronto's Green Roof Bylaw, established in 2009, requires new developments to include green roofs, which not only provide insulation but also reduce air pollution and enhance urban biodiversity (Toronto Municipal Code, Chapter 492, Green Roofs, 2021)

These projects demonstrate how NbS can bridge environmental and public health goals, reducing disease burdens while enhancing urban livability.

3. *Wetland Restoration for Flood Resilience.*
As climate change increases the frequency of hurricanes, floods, and extreme storms, hospitals and healthcare delivery systems must adapt to prevent infrastructure damage. Some HICs are using wetland restoration and natural flood barriers to protect healthcare facilities from extreme weather events.

   a. The Netherlands' "Room for the River" project has restored wetlands and floodplains near healthcare facilities, reducing flood risks and improving water quality (European Environment Agency, 2023).

   b. New Orleans, USA, is investing in coastal wetland restoration to buffer hospitals and residential areas from hurricane-induced flooding (Resilience and Sustainability-NOLA, 2025).

c. In Copenhagen, Denmark, hospitals are integrating blue-green infrastructure, such as rain gardens and permeable pavements, to manage stormwater and prevent infrastructure strain (C40 Cities Climate Leadership Group, 2015).

By leveraging natural water management systems, healthcare facilities in HICs can increase their climate resilience, reduce flood damage, and maintain operational stability during extreme weather events.

3.1.3 NbS Adoption in Low- and Middle-Income Countries (LMIC).

In low- and middle-income countries (LMICs), climate change has been identified to impact income equality at a regressive scale, where the poorest companies have the greatest economic and health struggles from climate change (Gilli et al., 2024). Adding to the disparity, healthcare infrastructure is often underfunded and overburdened, leaving opportunity for NbS to provide low-cost, scalable, and community-driven solutions to address public health challenges. The emphasis in LMICs is on nature-based resilience, as many regions face severe climate impacts, poor air quality, and limited access to healthcare.

1. *Passive Cooling and Green Building Materials.*
    In many tropical and arid LMICs, extreme heat and poor building ventilation contribute to heat-related illnesses and hospital energy inefficiencies. To counter this, NbS strategies include:
    a. Green roofs and vegetation-based shading, which passively cool hospital buildings in countries like India and Kenya, reducing energy demands for air conditioning (Malaeb et al., 2022).
    b. Traditional mud-brick and bamboo construction in rural hospitals, which provide natural insulation and temperature regulation, making healthcare facilities more comfortable and sustainable (ARCH, 2020; Carpio et al., 2024).

c. Rainwater harvesting systems, which are being integrated into hospitals in Bangladesh and Nigeria, ensuring access to clean water for sanitation and infection control (UNB, 2022; Lade et al., 2018).

These methods significantly lower energy costs and improve indoor air quality, making healthcare facilities more resilient and sustainable.

2. *Ecosystem-Based Water Filtration.*
    Access to clean water and sanitation remains a major challenge in LMICs, contributing to high rates of diarrheal diseases, cholera, and antimicrobial resistance (WHO, 2019). NbS approaches such as:
    a. Wetland restoration for water filtration in Uganda and Cambodia, improving water quality and reducing disease transmission (Namaalwa, S, et al., 2020; Liang, KM, et al., 2010).
    b. Bio-sand filtration and constructed wetlands near health-care facilities in sub-Saharan Africa, providing cost-effective water purification (Mekonnen et al., 2015).

These solutions reduce waterborne illnesses, ensuring cleaner, more sustainable water supplies for hospitals and communities.

3. *Community-Led Reforestation for Disease Control.*
    Deforestation increases the spread of vector-borne diseases such as malaria and dengue by disrupting natural predator-prey relationships. In response:
    a. Mangrove reforestation in Southeast Asia has been linked to lower mosquito populations, reducing malaria transmission (Reid et al., 2020).

b. Agroforestry projects in Latin America have helped restore biodiversity, indirectly decreasing zoonotic disease risks (Keesing et al., 2021).

By restoring natural habitats, communities can reduce disease vectors and improve long-term public health outcomes.

3.1.4 Conclusions.

In summary, Nature-based Solutions (NbS) provide a geographically adaptable, cost-effective, and sustainable approach to improving healthcare services' delivery and healthcare delivery system resilience in both HICs and LMICs. While HICs can integrate green infrastructure, urban forests, and climate-resilient hospitals, LMICs benefit from low-cost solutions like passive cooling, ecosystem-based water filtration, and reforestation for disease control. As climate change intensifies, NbS must become a core strategy in global healthcare planning, ensuring that medical systems remain sustainable, resilient, and environmentally responsible.

3.2 Integration of Ecosystems into Healthcare.

The integration of ecosystems into healthcare recognizes the profound connection between environmental health and human health, wellness, and well-being. From green spaces that enhance mental health to plant-based compounds integral to modern medicine, nature is essential in preventing and treating disease while supporting the overall health, wellness, well-being, and resilience of individuals and their communities.

As healthcare systems face increasing challenges such as chronic disease burdens, mental health crises, and antibiotic resistance, leveraging natural systems—through biophilic design, nature-based therapies, and ecosystem restoration—offers innovative, cost-effective solutions. Understanding how natural environments contribute to healing and resilience can help shape a more holistic, sustainable approach to healthcare delivery.

3.2.1 Ecosystems and Their Role in Nature-Based Solutions (NbS).

Ecosystems are the foundation of life on Earth, providing essential services such as climate regulation, water purification, and biodiversity support and conservation. As environmental challenges intensify due to Anthropocene climate change, urbanization, and resource depletion, nature-based solutions (NbS) have emerged as a sustainable approach to address these issues. By leveraging the resilience and functionality of ecosystems—such as forests, wetlands, and coastal habitats—NbS offer cost-effective and adaptive strategies for mitigating climate risks, enhancing biodiversity, and improving the human condition. Understanding the role of ecosystems in NbS is crucial for designing interventions that align ecological structures and processes with societal needs, preferences, and values, ensuring long-term environmental and economic benefits.

3.2.1.1 Ecosystems Types.

Ecosystems encompass the intricate web of living organisms and their physical environments, functioning through interdependent cycles and processes that sustain ecosystem services and biodiversity conservation. These systems include:

1. *Terrestrial ecosystems (forests, grasslands, tundra).*
2. *Aquatic ecosystems (rivers, lakes, wetlands).*
3. *Marine ecosystems (coastal zones, coral reefs, open oceans).*
4. *Soil ecosystems (microbial communities, nutrient cycles).*

Each of these ecosystems plays a critical role in regulating climate, purifying air and water, maintaining biodiversity, and supporting food production. The degradation of these ecosystems due to deforestation, pollution, resource depletion, urbanization, and climate change has led to significant environmental and socioeconomic challenges.

3.2.1.2 Nature-Based Solutions (NbS): Working with Ecosystems.

Previously introduced, nature-based solutions (NbS) are strategies that protect, restore, and sustainably manage ecosystems to address societal challenges, including climate change mitigation and adaptation, disaster risk reduction, food security, and human health (Cohen-Shacham et al., 2016). Rather than relying solely on engineered solutions (such as dams or seawalls), NbS integrate ecological structures and processes to enhance resilience, improve environmental quality, and provide cost-effective long-term benefits.

Some examples of NbS using ecosystems include:

1. *Forest Conservation & Reforestation (Terrestrial Systems).*
    a. Protecting and restoring forests helps mitigate climate change by sequestering greenhouse gases, reducing erosion, and maintaining biodiversity.
    b. Example: The Bonn Challenge, a global initiative aiming to restore 350 million hectares of degraded land by 2030 (IUCN, 2020).

2. *Wetlands Restoration (Aquatic Systems).*
    a. Wetlands act as natural sponges, absorbing excess floodwaters and improving water quality by filtering pollutants.
    b. Example: The Mississippi River Delta restoration in the U.S. enhances storm protection and restores habitats for fisheries (NOAA, 2021).

3. *Coastal Protection & Blue Carbon Ecosystems (Marine Systems).*
    a. Mangroves, seagrasses, and coral reefs protect shorelines from storm surges and erosion while sequestering large amounts of greenhouse gas.

b. Example: The Global Mangrove Alliance supports mangrove conservation as a tool for climate resilience (UNEP, 2020).

4. *Green Infrastructure & Urban NbS (Integrated Systems).*
   a. Green roofs, permeable pavements, and urban forests cool cities, absorb rainwater, and improve air quality.
   b. Example: Singapore's extensive use of green roofs and vertical gardens enhances urban resilience (Tan et al., 2014).

5. *Regenerative Agriculture & Soil Restoration (Soil Systems).*
   a. Techniques like no tillage seeding, cover cropping, agroforestry, and rotational grazing improve soil health, increase biodiversity, and enhance food security.
   b. Example: The "4 per 1000" Initiative aims to increase soil carbon stocks to mitigate climate change (Lal, 2020).

3.2.1.3 Benefits of Nature-Based Solutions (NbS).

Nature-based solutions (NbS) offer a range of advantages that make them a sustainable, efficient, and effective approach to addressing environmental, social, and economic challenges. These benefits primarily fall into three major categories:

1. *Cost-Effectiveness.*
   NbS provide a financially viable alternative to traditional engineered infrastructure by leveraging ecosystems that require lower maintenance and operational costs. For example, restoring wetlands for flood mitigation is often more cost-effective than constructing and maintaining concrete flood barriers, which require continuous upkeep and repairs. Additionally, green infrastructure—such as urban tree canopies and permeable surfaces—reduces expenses associated with stormwater management, cooling systems, and air pollution control, ultimately leading to long-term savings for municipalities and stakeholders.

2. *Multi-Functionality and Co-Benefits.*
    Unlike single-purpose engineered solutions, NbS deliver multiple benefits simultaneously. They enhance biodiversity conservation by creating or restoring habitats, contribute to climate change mitigation by sequestering greenhouse gases, and improve the publics' health by reducing pollution and providing spaces for recreation and mental well-being. For instance, urban green spaces not only support local ecosystems but also lower urban heat island effects, improve air quality, and promote physical activity, thereby reducing healthcare costs and improving overall community health, wellness, and well-being.

3. *Resilience and Adaptability.*
    One of the most significant advantages of NbS is their ability to evolve and adapt over time in response to environmental changes, making them inherently more resilient than rigid, human-made infrastructure. Unlike static concrete structures, which may degrade or become obsolete due to climate change, ecosystem solutions—such as restored mangroves, forests, or wetlands—can self-sustain, regenerate, and adjust to shifting conditions. By strengthening ecosystem resilience, NbS contribute to long-term sustainability, helping communities better withstand climate-related disasters such as floods, droughts, and extreme heat events.

    By integrating NbS into urban planning, infrastructure development, and climate adaptation strategies, societies can build more sustainable, cost-efficient, and resilient ecosystems that benefit both people and the environment.

3.2.1.4 Challenges of Implementing Nature-Based Solutions (NbS)

Despite their proven effectiveness in addressing environmental and societal challenges, nature-based solutions (NbS) face several barriers that hinder their widespread adoption and long-term success. These challenges primarily fall into three key areas: land use conflicts, scalability and financing, and social equity. Addressing these obstacles requires a holistic and comprehensive approach that integrates leadership, governance, community engagement, and policy alignment to ensure sustainable and inclusive implementation.

1. *Competing Land Use Interests.*
   A significant challenge in implementing NbS is balancing competing demands for land, particularly between conservation and economic activities such as agriculture, urban expansion, and industrial development. For instance, restoring wetlands for flood protection or biodiversity conservation may conflict with agricultural needs, while afforestation projects might compete with land designated for housing and infrastructure. Resolving these conflicts requires participatory land-use planning accounting for both ecological and economic priorities, ensuring that NbS complement rather than displace essential industries and livelihoods, and the repurposing of abandoned or demolished local infrastructure.

2. *Scaling and Financing Sustainable Projects.*
   While NbS offer cost-effective long-term benefits, securing upfront investment and long-term financial support remains a major hurdle. Many NbS projects struggle with funding due to uncertainties about return on investment, the complexity of quantifying ecosystem services, and a lack of financial incentives for private sector involvement. Additionally, NbS often require cross-sector collaboration, which can complicate funding mechanisms. Expanding financial models—such as public-private partnerships (PPP), green bonds, greenhouse gas markets, and payments for ecosystem services—can help mobilize resources for large-scale NbS implementation. Governments and interna-

tional organizations must also integrate NbS into climate adaptation funding frameworks to ensure consistent support.

3. *Ensuring Social Equity in NbS Planning and Implementation.*
The equitable distribution of NbS benefits remains a critical concern, as vulnerable and marginalized communities often lack access to green spaces, climate adaptation resources, and decision-making processes. NbS projects must prioritize inclusive planning to avoid exacerbating social inequalities. For example, urban greening initiatives should be designed to benefit all residents rather than contributing to "green gentrification," where lower-income communities are displaced due to rising property values. Meaningful community engagement and empowerment, participatory leadership and governance, and cultural humility are essential to ensuring that NbS address the needs of diverse populations and promote environmental justice.

Successfully integrating Nature-based Solutions (NbS) into mainstream environmental and developmental strategies requires strong leadership and governance frameworks that enforce public policies supporting sustainable land use and cross-sectoral collaboration. It also demands active community engagement to foster local participation, knowledge-sharing, and equitable decision-making, along with integrated public policies that align conservation efforts with economic development, urban planning, and public health initiatives.

By addressing these challenges through innovative financing, inclusive governance, and strategic policy integration, NbS can become a cornerstone of sustainable development, providing long-term benefits for both people and the planet.

In summary, ecosystems are fundamental to nature-based solutions, providing essential ecosystem services that support life, biodiversity, and resilience. By leveraging these ecosystems, NbS offer sustainable, cost-effective, and adaptive approaches to tackling global challenges while preserving biodiversity and enhancing the human condition.

3.2.2 Integrating Nature-based Solutions (NbS) into Hospital Design.

The integration of NbS into hospital design is transforming health-care facilities into environments that promote healing, enhance the patient and provider experience, and improve operational sustainability. By incorporating elements such as natural light, green spaces, sustainable water management, and biophilic architecture, hospitals can create a more therapeutic atmosphere while reducing their environmental impact. These nature-based approaches align with growing empirical evidence that exposure to natural elements can accelerate recovery, lower stress, and improve overall patient and staff experiences (Ulrich et al., 2008).

1. *Biophilic Design and Patient Recovery.*
   Biophilic design, which incorporates natural elements into built environments, has been shown to positively impact patient outcomes. Features such as large windows providing access to daylight, indoor gardens, and green walls contribute to lower stress levels, reduced pain perception, and shorter hospital stays. Studies have demonstrated that patients with views of nature require fewer pain medications and experience improved psychological wellness and well-being compared to those in windowless or artificially lit rooms (Ulrich, 1984).

2. *Healing Gardens and Outdoor Spaces.*
   Healing gardens and landscaped courtyards provide patients, visitors, and healthcare workers with access to nature, supporting mental health and reducing anxiety. These spaces offer quiet areas for relaxation, meditation, and social interaction, which can be particularly beneficial in pediatric, geriatric, and psychiatric care settings. For example, hospitals incorporating therapeutic gardens for dementia patients have reported improvements in cognitive function and reduced agitation (Marcus & Sachs, 2014).

3. *Natural Ventilation and Indoor Air Quality.*
Improving indoor air quality through natural ventilation and green infrastructure helps reduce airborne contaminants and hospital-acquired infections. Integrating plant-based air purification systems, such as indoor green walls, enhances air filtration while also adding aesthetic and psychological benefits. Additionally, designing hospital layouts to maximize cross-ventilation can improve airflow and reduce reliance on mechanical ventilation systems, leading to energy savings.

4. *Sustainable Water and Energy Management.*
Hospitals consume vast amounts of water and energy on a 24-7-365 operating cycle, making sustainability a crucial consideration in modern healthcare infrastructure design. Rainwater harvesting, green roofs, and permeable surfaces can enhance water conservation efforts, while energy-efficient building materials and solar panels can help reduce the hospital's GHG inventory. The inclusion of living walls and rooftop gardens can also improve insulation, leading to lower heating and cooling costs.

5. *Nature-Inspired Materials and Aesthetics.*
Using natural materials such as wood, stone, and eco-friendly textiles in hospital interiors fosters a warm, calming environment. The choice of color palettes inspired by natural landscapes, along with textures that mimic organic elements, can contribute to a sense of tranquility, reducing anxiety for both patients and healthcare providers. Research suggests that hospital rooms with nature-inspired designs lead to improved patient satisfaction and overall comfort (Huisman et al., 2012).

In conclusion, integrating nature-based solutions (NbS) into hospital infrastructure design enhances patient-centered care, fosters community engagement, improves energy efficiency, and promotes environmental sustainability. These strategies not only support better health outcomes but also contribute to the development of resilient and adaptable healthcare facilities for patients, their families, and staff. As the need for healing-oriented hospital environments continues to grow, NbS will play an increasingly essential role in shaping the future of healthcare spaces.

3.2.3 Integrating Ecosystem Services into Patient Care.

Integrating ecosystem services into patient care is an emerging approach that harnesses the healing power of nature to enhance recovery, improve health outcomes, and promote overall well-being. By leveraging natural environments, this strategy supports positive health, optimizes wellness, and contributes to a more holistic approach to patient care.

By incorporating natural elements such as green spaces, therapeutic landscapes, nature-based therapies, and biophilic design into healthcare delivery settings, providers can create holistic and comprehensive treatment environments that support both physical and mental health. Research increasingly highlights the positive impact of nature on reducing stress, improving immune function, and enhancing the patient experience, making it a valuable complement to conventional healthcare services delivery (Ulrich et al., 2008).

1. *Nature-Based Therapies and Rehabilitation.*
    Nature-based therapies, such as horticultural therapy, animal-assisted therapy, and ecotherapy, have been successfully integrated into patient care for various conditions. These therapies are particularly beneficial for mental health, neurorehabilitation, and chronic disease management. Horticultural therapy involves guided gardening activities that have been shown to reduce anxiety, enhance cognitive function, and improve physical rehabilitation (both gross and fine motor) in patients recovering from stroke or trauma (Marcus & Sachs, 2014). Animal-assisted therapy, including interactions with therapy dogs or equine-assisted

therapy, has been used to support patients with PTSD, autism, and chronic pain syndrome, improving emotional resilience and social engagement (Barker et al., 2016). Ecotherapy, like the Japanese practice of "Shinrin-yoku" or "forest bathing" which encourages outdoor activities in natural settings, has demonstrated benefits for patients with diseases of despair (i.e., depression, anxiety), and stress-related disorders, helping to regulate mood and promote relaxation (Bratman et al., 2015).

2. *Exposure to Green Spaces for Mental and Physical Health.*
Access to green spaces, whether through hospital gardens, parks, or community nature programs, plays a significant role in patient recovery and overall well-being. Studies show that patients who can see or interact with nature experience lower stress levels, reduced pain perception, and faster recovery times (Ulrich, 1984). Additionally, urban green spaces have been associated with lower rates of cardiovascular disease, improved respiratory health, and better overall public health outcomes (Twohig-Bennett & Jones, 2018).

3. *Natural Light and Circadian Health.*
Integrating natural light into patient care settings, such as hospitals and long-term care facilities, has been shown to regulate circadian rhythms, improve sleep quality, and reduce symptoms of depression and delirium in hospitalized patients (Figueiro et al., 2017). Patients exposed to adequate natural light during recovery also report lower pain levels and shorter hospital stays, emphasizing the importance of designing healthcare spaces that maximize daylight exposure.

4. *Nature-Inspired Nutrition and Holistic Healing.*
Many ecosystems are also integrated into patient care through nutrition and holistic medicine. Traditional healing practices, such as herbal medicine and plant-based diets, continue to

play a role in disease prevention and management. Nutritional programs that emphasize whole, organic foods derived from sustainable agriculture contribute to better patient outcomes, particularly in managing metabolic disorders, cardiovascular disease, and inflammatory conditions (Tilman & Clark, 2014). Whole, organic foods are nutrient-dense and minimally processed, cultivated or raised without synthetic pesticides, herbicides, genetically modified organisms (GMOs) engineered for size, taste, or shelf life, artificial additives, or routine antibiotics and hormones. These foods maintain their natural composition, preserving essential vitamins, minerals, fiber, and antioxidants that support overall health and well-being. Whole, organic foods include fresh fruits and vegetables, whole grains, legumes, nuts, seeds, and ethically sourced animal products produced through organic farming practices that prioritize environmental sustainability and soil health.

5. *Sensory Stimulation and Cognitive Health.*
    Natural systems contribute to sensory healing, particularly for patients with cognitive impairments, dementia, or neurodevelopmental disorders. Sensory gardens, water features, and soundscapes using nature sounds (e.g., birdsong, running water) have been shown to enhance cognitive function, reduce agitation, and improve emotional well-being in patients with Alzheimer's disease and related dementias (Whear et al., 2014).

In summary, integrating ecosystems into patient care provides a holistic, evidence-based and comprehensive approach to improving individual and community health, wellness, and well-being. Whether through nature-based therapies, green spaces, natural light, or sustainable nutrition, these strategies help create patient-centered environments that support healing, resilience, and long-term positive health. As healthcare delivery systems continue to evolve worldwide, incorporating nature into treatment plans and facility design will play an increasingly vital role in enhancing patient outcomes and promoting sustainable healthcare practices.

3.2.4 Integrating Ecosystem Services into Community-Oriented Primary Health Care (COPHC).

Community-Oriented Primary Health Care (COPHC) is a health-care approach that combines clinical medicine with essential public health functions and community-based services to address the broader determinants of health (i.e., social, economic, environmental, and commercial) within a community. Integrating ecosystem services into Community-Oriented Primary Health Care (COPHC) strengthens its impact by utilizing nature-based solutions (NbS) to improve health outcomes, prevent disease, and enhance community resilience. By incorporating green infrastructure, nature-based therapies, sustainable food systems, and environmental stewardship into primary health care models, COPHC can advance both individual and population health while promoting long-term, sustainable healthcare solutions.

1. *Green Infrastructure for COPHC.*
    Access to green spaces, such as parks, urban gardens, and tree-lined streets, plays a crucial role in reducing chronic disease risks and enhancing community wellness and well-being. Studies have shown that exposure to nature lowers rates of hypertension, obesity, and mental health disorders while encouraging physical activity and social cohesion (Twohig-Bennett & Jones, 2018). Integrating green infrastructure into COPHC involves:
    a. Establishing community gardens at health care centers to promote lifestyle change and social engagement.
    b. Encouraging *Park Prescription* programs, where physicians prescribe outdoor activities to manage conditions like anxiety, depression, and cardiovascular disease (James et al., 2016).

    c. Designing safe and walkable green neighborhoods that facilitate active transportation and reduce air pollution-related illnesses.

2. *Nature-Based Preventive and Therapeutic Interventions.*
Ecosystem services provide an array of preventive and therapeutic interventions that align with COPHC's goal of addressing health disparities at the community level. These include:
    a. *Horticultural Therapy:* Community gardens integrated with healthcare clinics offer opportunities for patients, families, and staff to engage in therapeutic gardening, which has been linked to reduced stress, improved mental health, and enhanced mobility for elderly and disabled individuals (Marcus & Sachs, 2014).
    b. *Ecotherapy and Green Care:* Nature-based mental health interventions, such as forest bathing and outdoor mindfulness programs, can help manage stress-related disorders, PTSD, and substance abuse recovery.
    c. *Traditional and Indigenous Healing Practices:* Many communities, particularly in rural and Indigenous populations, incorporate medicinal plants and natural remedies into primary care, blending traditional ecological knowledge with modern medical practices (WHO, 2019).

3. *Sustainable Food Systems and Nutritional Health.*
Food insecurity and poor nutrition are major public health concerns that COPHC aims to address. Integrating natural food systems into community health efforts ensures access to fresh, locally grown produce, which can reduce diet-related diseases such as diabetes and cardiovascular conditions. Key strategies include:
    a. Supporting farm-to-clinic programs (e.g., *Food as Medicine* programs) that provide patients with fresh produce and nutrition education.

b. Encouraging urban agriculture initiatives to promote food sovereignty and improve community resilience.

c. Partnering with local farmers' markets to expand access to healthy foods in underserved areas (Tilman & Clark, 2014).

4. *Environmental Health and Climate Resilience in COPHC.*
   Climate change and environmental degradation disproportionately affect marginalized and vulnerable populations, making environmental health a critical component of COPHC. Ecosystem services can be integrated to enhance community resilience by:

   a. Implementing green stormwater management (e.g., rain gardens, permeable surfaces) to prevent flooding and reduce vector-borne disease risks.

   b. Reducing heat-related illnesses through urban tree-planting initiatives, which lower temperatures and improve air quality.

   c. Educating communities on climate-adaptive health practices, such as sustainable water use and renewable energy adoption in healthcare settings.

In summary, integrating ecosystem services into Community-Oriented Primary Health Care (COPHC) strengthens the public's health by addressing health-related social needs, economic, and environmental determinants of health. By incorporating green infrastructure, nature-based therapies, sustainable food systems, and climate resilience strategies, COPHC can create healthier, more sustainable communities. As person-centered integrated systems of health frameworks continue evolving, embracing NbS within COPHC frameworks will be essential for improving health equity and environmental sustainability.

3.3 Nature-Based Solutions (NbS) Enhance Global Healthcare Delivery Systems' Resilience.

Nature-Based Solutions (NbS) provide innovative and sustainable strategies for enhancing the resilience of global healthcare delivery systems. By leveraging natural processes—such as ecosystem restoration, improved water management, and green infrastructure—NbS address environmental challenges while strengthening healthcare systems' ability to respond to crises. These solutions not only mitigate climate change impacts but also promote more efficient, cost-effective, and adaptive healthcare delivery. Integrating NbS into healthcare infrastructure and planning enables communities to build stronger, more resilient systems that can withstand disruptions and ensure the continuous provision of essential healthcare services.

3.3.1 Resilience Within the Context of NbS Enhancing Global Healthcare Delivery Systems.

In the context of NbS strengthening the resilience of global healthcare delivery systems, resilience refers to the ability of healthcare systems worldwide to effectively respond to, adapt to, and recover from a range of challenges, including climate-related events, environmental degradation, and health crises like pandemics or disease outbreaks.

NbS contribute to this resilience by offering sustainable and adaptive interventions that not only mitigate the adverse effects of environmental stressors but also enhance the overall capacity of healthcare delivery systems to continue providing essential healthcare services during periods of disruption. These solutions improve both the physical and social infrastructures that underpin healthcare services delivery and healthcare delivery systems, fostering more robust, equitable, and sustainable global healthcare services delivery.

1. *Physical Resilience through NbS.*

    Natural solutions such as green infrastructure, ecosystem resto-
    ration, and sustainable water management can directly enhance
    the physical resilience of healthcare facilities and their surround-
    ing communities. For example, hospitals and clinics can ben-
    efit from NbS like green roofs and rain gardens, which man-
    age stormwater, reduce flood risks, and lower the urban heat
    island effect (Foster et al., 2011). These solutions help protect
    healthcare facilities from extreme weather events, such as floods
    and heatwaves, ensuring continued access to care during envi-
    ronmental disruptions. By reducing dependence on conven-
    tional infrastructure, such as concrete flood barriers, NbS also
    reduce costs while providing long-term, sustainable benefits.
    Additionally, NbS improve air quality, a critical factor for both
    general public health and the effectiveness of healthcare delivery
    systems. Urban areas with more green spaces have been linked
    to better respiratory health outcomes and fewer hospital admis-
    sions for asthma and respiratory diseases (Tzoulas et al., 2007).
    As climate change exacerbates air pollution, these nature-based
    interventions are essential in ensuring that global healthcare
    delivery systems are not overwhelmed by respiratory illnesses.

2. *Social Resilience through NbS.*

    NbS also contribute to the social resilience of global healthcare
    delivery systems by fostering healthier communities and reduc-
    ing health disparities. For instance, community gardens and
    urban agriculture can improve access to nutritious food, espe-
    cially in underserved areas, while providing opportunities for
    social engagement and collective action. These initiatives can
    bolster upstream health-related social needs and public health
    by addressing food insecurity, which is linked to chronic dis-
    eases such as diabetes, hypertension, and malnutrition (Pretty
    et al., 2011). Furthermore, NbS can reduce health inequities
    by improving access to healthcare services in marginalized and

vulnerable populations. Green spaces and nature reserves within communities offer a therapeutic setting for individuals suffering from mental health issues, such as depression or PTSD, conditions that have been exacerbated by the stress of living in underserved, resource-poor areas (Bratman et al., 2015). By enhancing the mental and physical well-being of individuals, NbS ensure that communities are better equipped to withstand and recover from health crises, particularly those related to environmental and socioeconomic stressors.

3. *Adaptive Capacity of Global Healthcare Delivery Systems.*
    The integration of NbS into global healthcare delivery systems enhances their adaptive capacity—the ability to evolve in response to climate changing conditions. NbS promote sustainable healthcare services delivery by shifting from a reactive to a proactive approach, focusing on health prevention, health protection, health promotion, emergency preparedness and long-term sustainability. For example, the restoration of wetlands or mangroves not only provides protection against storm surges and flooding but also enhances biodiversity, which can play a critical role in preventing the spread of infectious diseases (Seddon et al., 2020). Additionally, NbS help build adaptive healthcare capacity by reducing healthcare costs and improving efficiency. Sustainable infrastructure such as renewable energy systems, water recycling, and green building designs can reduce operational costs for healthcare facilities, allowing resources to be redirected towards expanding care services, particularly in low-income areas (Elmqvist et al., 2015). This approach is essential as healthcare delivery systems globally face growing pressure to provide evidence-based, value-driven high-quality care with limited resources, particularly in the wake of climate-related disruptions.

Resilience in global healthcare delivery systems, enhanced through NbS, is a multi-faceted concept that includes physical, social, and adaptive dimensions. Nature-based solutions provide a sustainable, cost-effective means to strengthen healthcare services' delivery infrastructure and improve health outcomes. By incorporating NbS into healthcare planning and implementation, healthcare delivery systems can better withstand environmental challenges, promote health equity, and adapt to future global health needs, ensuring that care delivery remains robust in the face of crises.

3.3.2 Nature-Based Solutions (NbS) Reducing Global Healthcare Delivery System Operational and Management Costs.

As healthcare delivery systems around the world face increasing financial pressure from rising operational costs, particularly in the wake of climate change and the global health crises such as pandemics, integrating Nature-Based Solutions (NbS) offers a promising pathway to reducing both operational and management expenses. By harnessing ecosystems and ecosystem services, NbS not only help mitigate environmental impacts but also provide long-term financial savings and operational efficiencies for healthcare delivery systems. These cost-reducing benefits stem from improved infrastructure resilience, resource efficiency, and reduced health burdens, all of which contribute to lowering both direct and indirect healthcare costs.

1. *Cost-Effective Infrastructure and Energy Management.*
    NbS, such as green roofs, urban gardens, and natural stormwater management systems, are increasingly incorporated into healthcare facility designs to reduce the costs associated with energy consumption, water management, and routine maintenance. For instance, green roofs help to regulate indoor temperatures, reducing the need for air conditioning and heating, which can be major contributors to operational costs. Hospitals and healthcare facilities that incorporate these systems often see a decrease in energy consumption, leading to lower utility bills (Foster et al., 2011). Natural stormwater management systems, such as rain gardens

and permeable surfaces, reduce the need for expensive drainage infrastructure and flood management systems while also decreasing the risk of water-related disruptions. The cost-saving potential of such NbS is particularly evident in global regions prone to heavy rainfall and flooding, where traditional infrastructure is often inadequate and costly to maintain. By reducing flood risks and promoting water conservation, NbS help healthcare facilities avoid costly repairs, delays in healthcare services delivery, and operational disruptions due to water damage.

2. *Lowering Healthcare Costs by Reducing Disease Burden.*
   NbS not only enhance the physical and built environment of healthcare delivery settings but also improve public health practices, which, in turn, reduces the overall burden on global healthcare delivery systems. The integration of green spaces, parks, and urban forests into communities has been linked to improved mental and physical health outcomes, including reductions in stress, anxiety, and chronic diseases such as hypertension and cardiovascular disease (Tzoulas et al., 2007). By promoting physical activity, reducing pollution, and offering spaces for social interaction, NbS can lead to a reduction in the prevalence of lifestyle-related diseases that contribute to the rising costs of healthcare (Twohig-Bennett & Jones, 2018). Communities with access to green spaces experience fewer instances of respiratory and cardiovascular conditions, reducing the demand for medical services and hospital admissions. This leads to a long-term reduction in healthcare expenditures, as fewer patients require intensive treatments or hospital care, and overall public health improves.

3. *Proactive Healthcare and Chronic Disease Management.*
   NbS also play a role in proactive healthcare, a critical strategy for managing long-term healthcare costs. Nature-based interventions, such as outdoor exercise programs, green exercise initiatives, and ecotherapy, have been shown to improve mental

health, reduce stress, and promote physical activity (Bratman et al., 2015). By incorporating these activities into community-based primary health care (COPHC) programs, global healthcare delivery systems can reduce the prevalence of mental health disorders, obesity, and chronic conditions like diabetes and heart disease, ultimately lowering the costs associated with their treatment and management. In addition, community gardens and urban agriculture initiatives promote healthy eating habits and food sovereignty, leading to better nutrition and fewer diet-related diseases. These solutions address food insecurity, which is a major contributor to chronic conditions such as diabetes, hypertension, and obesity (Pretty et al., 2011). By supporting sustainable, localized food production, NbS help reduce the need for expensive medical interventions and improve the long-term health of populations.

4. *Enhanced Resilience to Climate Change and Health Emergencies.*
   One of the most significant cost-saving benefits of NbS in healthcare is the role they play in enhancing resilience to climate change and health emergencies. For example, NbS such as coastal wetlands and mangrove restoration help mitigate the impacts of extreme weather events like flooding, hurricanes, and storm surges, which are becoming more frequent due to climate change. These natural barriers reduce the need for costly emergency healthcare interventions and repairs to infrastructure following environmental disasters. Additionally, NbS such as urban heat island mitigation through increased tree cover and green infrastructure can help reduce the incidence of heat-related illnesses, which are particularly costly in regions experiencing more extreme heat waves. Hospitals can save on healthcare costs related to the treatment of heat stroke, dehydration, and heat-related cardiovascular and respiratory conditions by investing in urban greening and cooling initiatives (Harlan et al., 2006).

5. *Long-Term Economic and Environmental Sustainability.*
Beyond direct cost savings, NbS contribute to the economic sustainability of global healthcare delivery systems by reducing the environmental footprint of healthcare operations. The implementation of on-site renewable energy systems, such as solar panels or wind turbines, along with natural resource management practices like water conservation and waste reduction, can lower the long-term operating costs of healthcare facilities (Elmqvist et al., 2015). Furthermore, these strategies help healthcare delivery systems align with global sustainability goals, attracting funding and investment from environmentally conscious stakeholders, governments, and organizations focused on climate adaptation and resilience.

In summary, NbS offer a transformational opportunity to reduce operational and managerial costs in global healthcare delivery systems. By improving infrastructure efficiency, lowering healthcare-related expenditures, preventing disease, and increasing resilience to climate change, NbS create long-term cost savings while contributing to healthier communities. As healthcare systems worldwide continue to navigate the financial pressures of modern healthcare services delivery (e.g., advanced medical and information technology), integrating NbS will be an essential strategy for reducing costs, improving patient outcomes, and fostering sustainable, resilient clinical medicine and public health practices.

3.3.2 Nature-Based Solutions (NbS) Reducing Adverse Ecological Impacts on Global Healthcare Delivery Systems.

Nature-Based Solutions (NbS) offer a holistic and comprehensive approach to addressing the negative ecological impacts that strain global healthcare delivery systems. By leveraging ecosystems and ecosystem services, NbS can mitigate the adverse effects of environmental degradation, climate change, and loss of biodiversity, all of which have significant ramifications for public health and global healthcare delivery systems. The integration of NbS into healthcare services planning can help prevent or reduce health risks, lower healthcare costs, and enhance resilience to ecological threats, ultimately improving the efficiency and sustainability of healthcare systems worldwide.

1. *Mitigating Air Pollution and Improving Respiratory Health.*
   One of the most pressing ecological challenges affecting global healthcare delivery systems today is air pollution, which has profound impacts on the publics' health and wellness. The World Health Organization (WHO) attributes millions of premature deaths annually to air pollution, particularly from respiratory and cardiovascular diseases. By incorporating NbS such as urban forests, green spaces, and ecosystem buffers (i.e., prairies, wildflowers, pollinator corridors, shrublands, etc.) along highways and industrial zones, air quality can be significantly improved, while improving the health and resilience to surrounding ecosystems. Trees, grasses, shrubs, and natural air filters, absorb pollutants such as particulate matter (PM), nitrogen oxide (N2O), and sulfur dioxide (SO2), and release oxygen in return (Tzoulas et al., 2007). Urban greening such as pollinator gardens, pocket prairies, and trees along street corridors, for example, reduces the concentration of airborne pollutants, leading to fewer respiratory-related hospital admissions and reduced demand on healthcare services. Cities that have adopted such NbS have observed improvements in the public's health, including reductions in asthma rates and other chronic respiratory diseases, which in turn reduces the burden on healthcare systems (Dadvand et al., 2015). By integrating more green infrastructure into urban planning and healthcare facility design, we can help alleviate the adverse health impacts of air pollution and reduce the operational strain on healthcare systems.

2. *Addressing Heat Stress and Preventing Heat-Related Illnesses.*
Rising global temperatures, driven by climate change, have led to more frequent and severe heat waves, particularly in urban areas where the urban heat island effect exacerbates heat stress. This results in an increase in heat-related illnesses, such as heat stroke, dehydration, and exacerbations of cardiovascular and respiratory conditions. These heat-related illnesses contribute significantly to the burden on global healthcare delivery systems, with increased preventable emergency room visits and hospital admissions during heat events. NbS, such as urban tree canopy expansion, green roofs, and the restoration of native habitats, can significantly mitigate the urban heat island effect. Vegetation cools the environment by providing shade, releasing moisture through evapotranspiration, and reducing surface temperatures (Harlan et al., 2006). By reducing the intensity and frequency of heat-related health issues, NbS can lower the strain on emergency services and acute healthcare facilities during heatwaves. Additionally, incorporating green spaces in urban areas and healthcare facility designs can provide cooling areas for vulnerable populations, helping to prevent heat-related health crises before they occur.

3. *Reducing Flood Risks and Waterborne Diseases.*
Flooding is another ecological threat that negatively impacts global healthcare delivery systems, particularly in regions prone to extreme weather events or rising sea levels due to climate change. Floods disrupt healthcare facilities, displace communities, and contaminate water supplies, leading to the spread of waterborne diseases, such as cholera, dysentery, and typhoid fever. These diseases increase the demand for medical treatment and overwhelm healthcare resources, particularly in low- and middle-income countries (LMIC) (Watson et al., 2015). NbS, such as the restoration of wetlands, the establishment or

restoration of riparian buffers, and the creation of floodplain ecosystems, can reduce the frequency and severity of flooding events. Wetlands and mangroves act as natural sponges, absorbing excess rainfall and stormwater, thus preventing urban and rural areas from being inundated during heavy rains. By protecting and restoring these ecosystems, communities can reduce the likelihood of flooding, safeguard healthcare infrastructure, and mitigate the spread of waterborne diseases. Moreover, NbS offer a cost-effective alternative to traditional flood protection infrastructure, which can be expensive to build and maintain.

4. *Enhancing Biodiversity and Reducing Disease Spillover.*
    The loss of biodiversity, driven by habitat destruction, food-web interruptions, and climate change, can have significant consequences for public health practices. Disruptions to ecosystems increase the risk of disease spillover, where pathogens jump from animals to humans, often facilitated by changes in land use and habitat fragmentation. Diseases such as Zika, Ebola, SARS, and COVID-19 are examples of infectious diseases that have emerged or spread due to ecological imbalances and biodiversity loss (Keesing et al., 2017). NbS, including ecosystem restoration and conservation efforts to protect critical habitats and keystone wildlife species, are essential in maintaining global biodiversity and reducing the risk of zoonotic disease transmission. By protecting forests, wetlands, and other natural ecosystems, environmental health principles and practices (Battersby, 2022) help preserve the balance of ecosystem services that regulate disease dynamics. For example, preserving wetlands helps regulate mosquito populations, reducing the risk of malaria and other mosquito-borne diseases. In this way, NbS help mitigate the environmental factors that contribute to the emergence and spread of infectious diseases, ultimately reducing the burden on healthcare delivery systems.

5. *Promoting Healthy Communities and Reducing Healthcare Costs.*
    In addition to mitigating specific ecological threats, NbS have
    broader, long-term benefits for community health, wellness,
    and well-being, which can reduce the overall demand for health-
    care services delivery. Urban parks, green spaces, and commu-
    nity gardens encourage physical activity, social interaction,
    and stress reduction, all of which contribute to better mental
    and physical health outcomes. Stress reduction, improved car-
    diovascular health, and lower rates of chronic diseases such as
    obesity and diabetes translate to fewer hospital visits, chronic
    disease self-management, and a reduced strain on healthcare
    delivery systems (Pretty et al., 2011). Furthermore, NbS initia-
    tives often have co-benefits that align with public health goals,
    such as improved mental health, enhanced social cohesion, and
    increased opportunities for community engagement. These
    co-benefits reduce global healthcare delivery system expendi-
    tures by improving population health, preventing the onset of
    chronic conditions, and promoting community resilience in the
    face of ecological disruptions.

In summary, NbS is a powerful tool for reducing the adverse
ecological impacts on global healthcare delivery systems. By addressing environmental stressors such as air pollution, heat waves, flooding, biodiversity loss, and disease transmission, NbS not only protect healthcare infrastructure but also promote better public health
outcomes. The integration of NbS into healthcare services planning
and implementation can help prevent the escalation of health risks,
reduce the need for expensive interventions, and ensure that healthcare systems are more resilient and sustainable in the face of future
environmental challenges.

3.4 References.

1. ARCH 20. (2020). *Innovative Adaptation of Traditional Building Materials for the Modern Era.*
2. Barker, S. B., Knisely, J. S., McCain, N. L., & Best, A. M. (2016). Measuring stress and immune response in healthcare professionals following interaction with a therapy dog: A pilot study. *Psychological Reports,* 118(1), 180-198.
3. Battersby, S. (2022). Historical context, philosophy, and principles of Environmental Health. *In* Clay's Handbook of Environmental Health. Routledge: London.
4. Bratman, G. N., Hamilton, J. P., Hahn, K. S., Daily, G. C., & Gross, J. J. (2015). Nature experience reduces rumination and subgenual prefrontal cortex activation. *Proceedings of the National Academy of Sciences,* 112(28), 8567-8572.
5. C40 Cities Climate Leadership Group. (2015). *Cities 100: Copenhagen-Green Infrastructure Prevents Flooding.*
6. Carpio, R., Valarezo, F., Aguirre-Maldonado, E., & Balcázar-Arciniega, C. (2024). Influence of the Rustic Bamboo Envelope Construction Technique on the Thermal Performance of Vernacular Housing in the Ecuadorian Coastal Region: The Case of El Carmen-Manabí. *Buildings, 14(11), 3368.*
7. Cohen-Shacham, E., Walters, G., Janzen, C., & Maginnis, S. (2016). Nature-based solutions to address global societal challenges. *IUCN.*
8. Dadvand, P., de Nazelle, A., & Basagaña, X. (2015). The effects of urban green space on the health of the population: A systematic review. *Environmental Health Perspectives*, 123(7), 597-609.
9. Dzhambov, A. M., Hartig, T., Markevych, I., Tilov, B., & Dimitrova, D. (2021). Urban residential greenspace and mental health in youth: Different approaches to testing multiple pathways yield inconsistent results. *Environmental Research, 193,* 110469.
10. Ellen MacArthur Foundation (EMF). *We Need to talk about renewables.* Netherlands, EMF.

11. Elmqvist, T., Bai, X., & Frantzeskaki, N. (2015). *Urbanization, biodiversity and ecosystem services: Challenges and opportunities.* Springer.

12. European Environment Agency (EEA). (2023). *Interview-The Dutch make room for the river.*

13. Figueiro, M. G., Plitnick, B., Rea, M. S., Gras, L. Z., & Rea, M. S. (2017). Lighting intervention to improve sleep in institutionalized older adults with Alzheimer's disease: A pilot study. *Clinical Interventions in Aging*, 12, 1931-1940.

14. Foster, J., Lowe, A., & Soper, J. (2011). The role of green infrastructure in enhancing urban resilience. *Journal of Urban Planning and Development*, 137(3), 166-176.

15. Gilli, M., Calcaterra, M., Emmerling, J. & Granella, F. (2024). *Climate change impacts on the within-country income distribution.* Journal of Environmental Economics and Management, 127, 1-12.

16. Grün Berlin GmbH. (2025). *Development of subsections "Berlin Wall green corridor".*

17. Harlan, S. L., Brazel, A. J., Prashad, L., Minner, J., & Stehr, J. (2006). *Neighborhood effects on heat deaths: Social and environmental predictors of vulnerability in Maricopa County, Arizona.* Environmental Health Perspectives, 114(3), 460-467.

18. Hoffman, J. S., Shandas, V., & Pendleton, N. (2020). The effects of historical housing policies on resident exposure to intra-urban heat. *Climate, 8*(1), 12.

19. Huisman, E. R., Morales, E., Van Hoof, J., & Kort, H. S. (2012). Healing environment: A review of the impact of physical environmental factors on users. *Building and Environment*, 58, 70-80.

20. International Union for Conservation of Nature (IUCN). (2020). *Global standard for nature-based solutions: A user-friendly framework for the verification, design, and scaling up of NbS.*

21. James, P., Banay, R. F., Hart, J. E., & Laden, F. (2016). A review of the health benefits of greenness. *Current Epidemiology Reports*, 3(2), 131-142.

22. Jimenez, M. P., DeVille, N. V., Elliott, E. G., Schiff, J. E., Wilt, G. E., Hart, J. E., & James, P. (2021). Associations

between nature exposure and health: A review of the evidence. *International Journal of Environmental Research and Public Health, 18*(9), 4790.

23. Keesing, F., Belden, L., & Daszak, P. (2017). Impacts of biodiversity on the emergence and transmission of infectious diseases. *Nature*, 472(7343), 51-58.

24. Keesing, F., & Ostfeld, R. S. (2021). Impacts of biodiversity and biodiversity loss on zoonotic diseases. *Proceedings of the National Academy of Sciences of the United States of America, 118*(17), e2023540118.

25. Lade, Omolara & Oloke, David. 2018. Performance Evaluation of a Rainwater Harvesting System: A Case Study of University College Hospital, Ibadan City, Nigeria. *Current Journal of Applied Science and Technology,* 25 (5), 1-14.

26. Lal, R. (2020). Regenerative agriculture for food and climate. *Journal of Soil and Water Conservation*, 75(5), 123A-126A.

27. Liang, K., M. Sobsey and C. Stauber. (2010). Improving Household Drinking Water Quality: Use of Biosand Filter in Cambodia. *Water and Sanitation Program Field Note, Water and Sanitation Program.*

28. Malaeb, L., Ghaddar, A., & Faour, G. (2022). Sustainable building practices in low-resource healthcare settings: A systematic review. *Sustainability, 14*(5), 2774.

29. Marcus, C. C., & Sachs, N. A. (2014). *Therapeutic landscapes: An evidence-based approach to designing healing gardens and restorative outdoor spaces.* John Wiley & Sons.

30. Mekonnen, A., et al. (2015). Wastewater treatment performance efficiency of constructed wetlands in African countries: A review. *Water science and technology: a Journal of the International Association on Water Pollution Research.* 71. 1-8. 10.2166/wst.2014.483.

31. Namaalwa S, van Dam AA, Gettel GM, et al. (2020) The Impact of Wastewater Discharge and Agriculture on Water Quality and

Nutrient Retention of Namatala Wetland, Eastern Uganda. Front. Environ. Sci. 8:148.

32. National Oceanic and Atmospheric Administration (NOAA). (2021). *Mississippi River Delta restoration projects.*

33. Pretty, J., Peacock, J., Hine, R., Sellens, M., & South, N. (2011). The health benefits of green exercise: A research agenda. *Environmental Health Perspectives, 111*(5), 701-707.

34. Reid, C. E., Clougherty, J. E., Shmool, J. L. C., & Kubzansky, L. D. (2020). Is all urban green space the same? A comparison of the health benefits of trees and grass in New York City. *Environmental Research, 185,* 109379.

35. Resilience and Sustainability.NOLA.Gov. (2025). *Bayou Bienvenue Wetland Restoration.*

36. Seddon, N., Chausson, A., Berry, P., Girardin, C., Smith, A., & Turner, B. (2020). Understanding the value and limits of nature-based solutions to climate change and other global challenges. *Philosophical Transactions of the Royal Society B, 375*(1794).

37. Seddon, N., Mace, G., & Cummings, R. (2020). Nature-based solutions to climate change and environmental degradation: The role of ecosystem restoration in health and well-being. *Global Environmental Change, 61,* 102073.

38. Tan, P. Y., Wang, J., & Sia, A. (2014). Perspectives on greening cities: The case of Singapore. *Urban Forestry & Urban Greening, 13*(3), 771-781.

39. Tilman, D., & Clark, M. (2014). Global diets link environmental sustainability and human health. *Nature, 515*(7528), 518-522.

40. Toronto Municipal Code. (2021). *Chapter 492, Green Roofs.*

41. Twohig-Bennett, C., & Jones, A. (2018). The health benefits of the great outdoors: A systematic review and meta-analysis of greenspace exposure and health outcomes. *Environmental Research, 166,* 628-637.

42. Tzoulas, K., Korpela, K., Venn, S., Yli-Pelkonen, V., Kaźmierczak, A., & Niemelä, J. (2007). Promoting ecosystem and human

health in urban areas using green infrastructure: A literature review. *Landscape and Urban Planning*, 81(3), 167-178.

43. Ulrich, R. S. (1984). View through a window may influence recovery from surgery. *Science*, 224(4647), 420-421.

44. Ulrich, R. S., Simons, R. F., Losito, B. D., Fiorito, E., Miles, M. A., & Zelson, M. (1991). Stress recovery during exposure to natural and urban environments. *Journal of Environmental Psychology*, 11(3), 201-230.

45. United News of Bangladesh (UNB). (2022). Climate Conscious Architecture: Bangladesh's Rainwater-harvesting Hospital Wins International Award.

46. Ulrich, R. S., Zimring, C., Zhu, X., DuBose, J., Seo, H. B., Choi, Y. S., Quan, X., & Joseph, A. (2008). A review of the research literature on evidence-based healthcare design. *Health Environments Research & Design Journal*, 1(3), 61-125.

47. United Nations Environment Programme (UNEP). (2020). *The Global Mangrove Alliance: A decade for ecosystem restoration.*

48. Watson, J., Roper, L., & Wilson, S. (2015). Flooding and its health impacts: An assessment of healthcare needs and infrastructure resilience. *Environmental Health Perspectives*, 123(5), 525-531.

49. Whear, R., Coon, J. T., Bethel, A., Abbott, R., Stein, K., & Garside, R. (2014). What is the impact of using outdoor spaces such as gardens on the physical and mental well-being of those with dementia? A systematic review of quantitative and qualitative evidence. *Journal of the American Medical Directors Association*, 15(10), 697-705.

50. World Health Organization (WHO). (2019). *WHO traditional medicine strategy: 2014-2023.* Geneva: WHO.

51. World Health Organization (WHO). (2019). Water, sanitation, and hygiene in healthcare facilities: Status and trends. *WHO Report.*

4.0

Sustainability in Global Healthcare Delivery Systems.

4.1 Guiding Healthcare Organizations World-wide Toward Sustainable Practices.

Healthcare organizations worldwide must adopt sustainable practices to balance quality patient care with environmental and financial responsibility. By integrating energy-efficient infrastructure, reducing medical waste, and optimizing resource allocation, institutions can lower costs and minimize their ecological impacts and dependencies. Additionally, implementing digital health solutions and value-based care models promotes efficiency while enhancing patient outcomes. Global collaboration and policy-driven incentives further support the transition toward sustainability, ensuring that healthcare remains accessible and resilient for future generations (Eckelman & Sherman, 2016; Pichler et al., 2019; Watts et al., 2021).

4.1.1 Principles of Sustainability in Global Healthcare Delivery Systems and Healthcare Services Delivery.

The principles of sustainability (i.e., Environmental, Social, and Governance pillars (ESG)) serve as a framework for responsible, sustainable, and equitable healthcare delivery systems worldwide. As healthcare organizations face increasing challenges related to

Anthropocene climate change, social disparities, and ethical gover-nance, integrating principles of sustainability is essential to ensuring high-quality healthcare services delivered in both an environmentally responsible and socially inclusive manner while adhering to robust governance standards. This approach fosters resilience, efficiency, and long-term sustainability in healthcare service delivery.

4.1.1.1 Environmental Stewardship: Sustainable Healthcare Operations.

The environmental component of sustainability in healthcare focuses on reducing the sector's ecological dependencies, impacts, risks, and opportunities (DIROs) through environmental stewardship, energy efficiency, and waste reduction. The healthcare industry is a significant contributor to global greenhouse gas (GHG) emissions, accounting for nearly *4.4% of net global emissions* (Karliner et al., 2020). Hospitals, pharmaceutical companies, and medical supply chains contribute to pollution, excessive energy consumption, and waste production.

1. *Decarbonization.*

   Healthcare facilities, operating 24-7-365, consume large amounts of energy for medical equipment, heating, ventilation, and air conditioning (HVAC) systems, and lighting. Transitioning to renewable energy sources, both onsite or renewable energy certif-icates (RECs) such as solar, wind, nuclear and geothermal power can significantly reduce hospitals' GHG emissions (Eckelman & Sherman, 2016). Additionally, sustainable hospital design incorporating LEED-certified (i.e., Leadership in Energy and Environmental Design) building materials and energy-efficient infrastructure may improve operational efficiency (Bilec et al., 2019).

2. *Sustainable Medical Supply Chains.*
   Pharmaceutical production and medical supply chain logistics contribute to pollution through energy-intensive manufacturing, packaging waste, and emissions from transportation. Green procurement policies—such as prioritizing biodegradable packaging, reducing single-use plastics, and adopting closed-loop recycling systems—can mitigate environmental harm (Practice Greenhealth, 2021). Furthermore, adopting regional and local (when available) supply chain strategies can reduce GHG emissions associated with long-distance transportation of medical products (Pichler et al., 2019).

3. *Waste Reduction and Circular Economy.*
   Medical waste, including hazardous materials, single-use plastics, and pharmaceutical residues, pose adverse environmental and health risks. Strategies such as waste segregation, autoclaving infectious waste, reprocessing medical devices, and reducing unnecessary diagnostic testing can lower waste production (WHO, 2020). Implementing circular economy principles—where medical equipment is refurbished and redistributed—enhances resource efficiency (McGain & Naylor, 2014).

4. *Digital Healthcare Solutions and Telemedicine.*
   Advancements in digital health can also contribute to environmental sustainability. Telemedicine, remote patient monitoring, and AI-driven diagnostics reduce the need for patient travel, thereby cutting transportation-related emissions (Purohit et al., 2021). Moreover, digitizing health records reduces paper consumption and enhances administrative efficiency (Bodenheimer & Sinsky, 2014).

4.1.1.2 Social Responsibility: Equitable Access to Care, High-Impact Leadership, and Workforce Wellness.

The social aspect of the principles of sustainability focuses on equitable access to healthcare, workforce wellness and well-being, diversity, and patient-centered care. The COVID-19 pandemic exposed significant global disparities in healthcare access, emphasizing the need for stronger social policies to address inequities in healthcare services delivery (Basu et al., 2022).

1. *Health Equity and Access to Care.*
   Healthcare equity ensures that all individuals—regardless of socioeconomic status, race, gender, or geographic location—have access to high-quality medical care. The World Health Organization's (WHO) Universal Health Coverage (UHC) initiative advocates for healthy public policies that remove financial barriers to healthcare services, expand integrated person-centered primary health care (IPC-PHC) services, and integrate community-based health programs focused on upstream health-related social needs (HRSNs) (WHO, 2021). Investments in mobile health clinics, community health workers, and culturally-competent and integrated person-centered systems of health can reduce health disparities and improve outcomes in underserved populations (Fiscella & Sanders, 2016).

2. *High-Impact Leadership in Global Healthcare Systems.*
   High-impact leadership (Swensen et al., 2013) in global healthcare delivery systems is essential for driving sustainable improvements in healthcare access, quality, and equity. Effective leaders in this space must navigate complex challenges, including resource constraints, political dynamics, and shifting disease burdens, while fostering innovation and resilience. High-impact leadership emphasizes systems thinking, adaptive capacity, and cross-sector collaboration, ensuring that health initiatives are

scalable and sustainable (Frenk et al., 2010). Leaders must also embrace transformational leadership styles, promoting a shared vision that aligns with universal health coverage (UHC) and sustainable development goals (SDGs) (Kickbusch & Gleicher, 2012). By prioritizing evidence-based decision-making and inclusive governance, leaders can build resilient health systems that withstand global crises, such as pandemics and climate-related health threats (Kruk et al., 2018). Furthermore, high-impact leaders in global healthcare systems must cultivate cultural humility and community engagement, ensuring that healthcare policies and interventions are modified to the unique needs of diverse populations (Bhutta et al., 2017). Investing in leadership development programs and mentorship opportunities for healthcare professionals—particularly in low- and middle-income countries (LMICs)—is crucial for building the next generation of global healthcare system leaders (WHO, 2021). Additionally, leveraging digital health technologies and data analytics enhances decision-making, allowing for more efficient resource allocation and improved patient outcomes (Rottingen et al., 2017). By fostering ethical, transparent, and patient-centered leadership, global healthcare leaders can drive sustainable transformation and equity in healthcare delivery systems worldwide.

3. *Healthcare Workforce Wellness and Well-Being.*
    Healthcare professionals face high levels of fatigue, burnout, and diseases of despair (e.g., depression, anxiety, and suicide) due to workload pressures, long hours, and emotional distress. Addressing workforce wellness and well-being through mental health support programs, flexible work schedules, and workload management strategies can improve performance, retention, and recruitment (Shanafelt et al., 2017). Organizations that implement supportive work environments and professional development opportunities contribute to a resilient and engaged healthcare workforce (National Academy of Medicine, 2019).

4.1.1.3 Governance in Sustainable Healthcare: Ethical, Transparent, and Accountable.

Governance in healthcare focuses on ethical leadership, regulatory compliance, data security, and responsible financial management. Strong governance structures (e.g., Board of Directors, Trustees, etc.) ensure that healthcare organizations operate transparently, ethically, truthfully, and in alignment with both traditional medicine and public health goals (Hofmarcher et al., 2021).

1. *Regulatory Compliance and Ethical Standards.*
   Governance ensures adherence to regulatory frameworks, including patient safety protocols, financial transparency, and anti-corruption policies. Ethical healthcare governance also includes eliminating conflicts of interest, preventing fraud, waste, and abuse, and ensuring clinical trials are conducted transparently, responsibly and ethically (OECD, 2020).

2. *Data Security and Digital Health Governance.*
   With the rapid adoption of digital health technologies, protecting patient data privacy and adaptive cybersecurity has become a critical governance issue. Healthcare delivery systems and organizations must implement robust cybersecurity measures, comply with data protection regulations (e.g., GDPR, HIPAA), and enhance interoperability between electronic health records (EHRs) to improve clinical efficiency while maintaining privacy for patient's healthcare and personal identifiable information (Dunn et al., 2019).

3. *Sustainable Financial Management.*
   Financial sustainability in healthcare governance ensures that limited healthcare resources are allocated efficiently without compromising patient care quality. Value-based care models— which focus on patient outcomes rather than service volume—

promote cost-effective care (Porter & Lee, 2013). Governments and healthcare organizations must adopt transparent financial reporting, fair pricing strategies, and responsible investment in public health infrastructure (WHO, 2021).

4.1.1.4 Conclusion.

The implementation of the principles of sustainability in global healthcare delivery systems is essential for ensuring responsible, equitable, and ethical healthcare services delivery worldwide. By prioritizing environmental stewardship, social responsibility, and strong governance, healthcare organizations can optimize patient-centered care, reduce ecological impact, and build resilient and sustainable healthcare infrastructures for future generations. Policymakers, healthcare leaders, and the medical-industrial complex must collaborate to integrate strategies that foster long-term sustainability, health equity, and responsible governance.

4.1.2 Designing and Implementing Sustainable Practices in Global Healthcare Services Delivery.

Sustainable healthcare service delivery is essential for ensuring that global healthcare delivery systems and the healthcare services delivered remain effective, resilient, and equitable in the face of global challenges such as Anthropocene climate change, healthcare resource limitations, aging world population, institutional bias, and rising healthcare costs. Designing and implementing sustainable practices in global healthcare services delivery involves integrating environmental responsibility and stewardship, economic efficiency, and social equity into sustainable healthcare systems operations and management. This approach enhances patient care, reduces ecological and ecosystem impact, and strengthens healthcare systems' infrastructure for long-term resilience and sustainability (Eckelman & Sherman,

2016). Sustainable global healthcare systems prioritize resource efficiency, low-GHG emission operations, waste reduction, and resilient infrastructure design while also fostering equitable access and ethical governance (Karliner et al., 2020).

4.1.2.1 Environmental Responsibility and Stewardship in Healthcare Services Delivery.

Healthcare facilities world-wide have a significant environmental footprint, with hospitals and medical institutions among the most resource-intensive sectors globally. Previously mentioned, the healthcare industry contributes approximately 4.4% of global GHG emissions, primarily due to energy consumption, transportation, and waste generation (Karliner et al., 2020). Implementing sustainable practices in healthcare operations is critical to reducing this environmental impact while maintaining high-quality patient care and achieving positive health outcomes.

1. *Energy Efficiency and Renewable Energy.*
   Healthcare facilities consume large amounts of energy for advanced medical technology and equipment, heating, ventilation, and air conditioning (HVAC) systems, and lighting. Transitioning to renewable energy sources, onsite or RECs, such as solar, wind, nuclear, and geothermal power can significantly lower healthcare-related emissions (Eckelman & Sherman, 2016). The adoption of energy-efficient hospital designs, LED lighting, and smart HVAC systems also contributes to lower energy consumption and reduced costs (Bilec et al., 2019). Sustainable healthcare infrastructure, such as LEED-certified (Leadership in Energy and Environmental Design) buildings, has been shown to improve operational efficiency and environmental performance (Allen et al., 2015).

2. *Sustainable Medical Waste Management.*
Medical waste, including hazardous materials, single-use plastics, and pharmaceutical residues, poses significant environmental and public health risks. Implementing comprehensive waste management strategies, such as waste segregation, reprocessing of medical devices, and autoclaving infectious waste, can help reduce healthcare waste (WHO, 2020). Circular economy approaches—where medical equipment and supplies are refurbished, repurposed, and reused—also contribute to sustainability (McGain & Naylor, 2014). Furthermore, reducing unnecessary medical and surgical procedures and overprescription of pharmaceuticals can lower both costs and environmental harm (Practice Greenhealth, 2021).

3. *Greenhouse Gas Reduction and Digital Healthcare Solutions.*
Digital health solutions, such as telemedicine, remote patient monitoring, and AI-driven diagnostics, offer sustainable alternatives to traditional healthcare services delivery. By reducing the need for patient travel and hospital visits, these technologies help lower transportation-related emissions and healthcare costs (Purohit et al., 2021). Additionally, the transition to electronic health records (EHRs) minimizes paper waste and improves administrative efficiency (Bodenheimer & Sinsky, 2014).

4.1.2.2 Economic and Social Sustainability in Healthcare Services Delivery

Sustainable healthcare services delivery extends beyond safe environmental practices to include economic viability and social responsibility. A well-designed healthcare delivery system ensures financial sustainability, workforce well-being, and equitable access to healthcare services (Porter & Lee, 2013).

1. *Financial Sustainability and Value-Based Care.*
    Traditional healthcare models often prioritize service volume over patient outcomes, leading to inefficiencies, waste, and rising costs. Transitioning to value-based alternative payment models, which focus on proactive care, patient-centered treatment, and outcome-driven reimbursement, enhances both economic sustainability and patient well-being (Porter & Lee, 2013). By minimizing preventable hospital readmissions, optimizing the allocation of limited resources, and integrating digital health solutions, healthcare organizations can enhance service quality while achieving sustainable cost savings (OECD, 2020).

2. *Workforce Sustainability, Wellness, and Well-Being.*
    Healthcare professionals experience high levels of fatigue and burnout due to workload pressures, long hours, emotional stress, and diseases of despair (i.e., depression, anxiety, and suicide). Ensuring workforce sustainability involves investing in mental health support programs, flexible work arrangements, and leadership development (Shanafelt et al., 2017). Organizations that prioritize employee wellness and well-being and professional growth and development contribute to a more resilient, sustainable, and engaged workforce, ultimately improving patient outcomes (National Academy of Medicine, 2019).

3. *Health Equity and Access to Care.*
    Ensuring universal health coverage (UHC) and equitable access to healthcare services is a fundamental pillar of sustainable global healthcare delivery systems. Health disparities, particularly in low- and middle-income countries (LMICs), can be mitigated by expanding primary health care services, strengthening community health programs, and integrating culturally competent models of care (WHO, 2021). Mobile health clinics and public-private partnerships play a vital role in extending care to

underserved, marginalized, and vulnerable populations (Fiscella & Sanders, 2016).

4.1.2.3 Governance and Policy Frameworks for Sustainable Healthcare Services Delivery.

Sustainability in global healthcare delivery requires robust governance frameworks, transparent healthy public policies, and strong regulatory compliance. Governments, healthcare institutions, and global health organizations must collaborate to implement healthy public policies that support sustainability while ensuring high-quality care delivery (Hofmarcher et al., 2021).

1. *Regulatory Compliance and Ethical Governance.*
   Strong governance ensures compliance with environmental regulations, ethical financial practices, and data security standards. Ethical healthcare governance includes eliminating corruption, promoting transparency in decision-making, and ensuring equitable resource distribution (OECD, 2020). Healthy public policies promoting environmental responsibility, mandatory stewardship reporting, and healthcare GHG inventory assessments foster accountability and drive meaningful progress (Karliner et al., 2020).

2. *Investment in Sustainable Healthcare Innovations.*
   Governments and healthcare organizations world-wide must invest in research, technology, and infrastructure to support sustainable healthcare service delivery models. Funding for renewable energy projects, digital health technologies, and climate-resilient healthcare systems ensures long-term sustainability and resilience (WHO, 2021). Additionally, financial incentives for hospitals and clinics that adopt green practices can accelerate the transition to sustainable healthcare service delivery (Pichler et al., 2019).

4.1.2.4 Conclusion.

Designing and implementing sustainable practices in healthcare services delivery is critical for ensuring long-term resilience, cost efficiency, and equitable patient care. By integrating environmental stewardship, economic efficiency, and social responsibility, global healthcare delivery systems can reduce their ecological impacts and dependencies while improving operational efficiency and patient outcomes. Key strategies include renewable energy adoption, waste reduction, digital health transformation, value-based alternative payment models, and workforce well-being initiatives. Strong governance frameworks and healthy public policy support for health-related social needs (HRSNs) and other determinants of health further ensure that sustainability remains a core principle of healthcare services delivery worldwide. As healthcare service delivery continues to evolve around-the-world, prioritizing sustainability and resilience will be essential for building robust, patient-centered, and environmentally responsible healthcare delivery systems.

4.2 Global Healthcare System Policies and Operations Incorporating Sustainability Principles.

The integration of Sustainability's Environmental, Social, and Governance (ESG) principles into global healthcare system policies and operations is critical for building resilient, sustainable, and equitable healthcare systems. ESG-driven sustainable healthcare policies focus on reducing environmental impacts, promoting social equity, and ensuring strong governance structures. These principles align with the United Nations Sustainable Development Goals (SDGs) and global health equity initiatives, shaping healthcare policies that enhance patient care, operational efficiency, and environmental responsibility (WHO, 2021).

4.2.1 Environmental Sustainability in Healthcare Policies and Operations.

Healthcare systems contribute significantly to global environmental challenges, including greenhouse gas (GHG) emissions, medical waste, and resource overconsumption. Environmental sustainability policies aim to mitigate healthcare's environmental dependencies, impacts, risks, and opportunities (DIROs)through green infrastructure, energy efficiency, sustainable procurement, and waste management strategies.

1. *Decarbonization of Healthcare Operations.*
   Many global healthcare delivery systems are adopting GHG-reducing energy solutions, such as renewable energy sources, to reduce their reliance on fossil fuels. Policies encouraging solar and wind energy adoption in hospitals, as seen in initiatives by the UK's National Health Service (NHS) and Health Care Without Harm (HCWH), have demonstrated significant greenhouse gas reductions (Karliner et al., 2020). Additionally, implementing energy-efficient hospital designs, green building certifications, and smart healthcare facilities has been shown to lower operational costs while improving patient outcomes (Allen et al., 2015).

2. *Sustainable Procurement and Supply Chain Management.*
   Sustainable environmentally-focused healthcare policies emphasize green procurement strategies, ensuring that medical products, pharmaceuticals, and hospital supplies meet environmental and ethical sourcing criteria. The European Union's Green Public Procurement (GPP) program sets standards for eco-friendly medical products and sustainable hospital operations (OECD, 2020). Additionally, the adoption of circular economy models, where medical devices and personal protective equipment (PPE) are refurbished and reused, is reducing healthcare waste while maintaining safety and quality standards (McGain et al., 2019).

3. *Waste Management and Water Conservation.*
   Global healthcare delivery systems are implementing stringent waste segregation and disposal policies to minimize the impact of hazardous and non-hazardous medical waste. The World Health Organization's (WHO) guidelines on healthcare waste management advocate for waste-to-energy technologies, recycling programs, and the reduction of single-use plastics in hospitals (WHO, 2020). Additionally, environmentally-focused policies promoting water conservation in healthcare facilities, such as low-flow fixtures and wastewater treatment solutions, are essential for ensuring sustainability in water-scarce regions (Pichler et al., 2019).

4.2.2 Sustainable Social Responsibility in Global Healthcare Systems.

Social sustainability in healthcare policies ensures health equity, workforce wellness and well-being, and community engagement. Sustainable socially-responsible healthcare delivery systems prioritize universal health coverage (UHC), patient-centered care, and diversity and inclusion initiatives to promote accessible and equitable healthcare for all populations.

1. *Health Equity and Universal Health Coverage (UHC).*
   Ensuring that healthcare services are accessible to all, regardless of socio-economic status, is a foundational cornerstone for socially-responsible sustainable practice. Countries with strong universal health coverage (UHC) policies, such as those in Scandinavia and Canada, have demonstrated improved health outcomes and reduced disparities (WHO, 2021). The Global Health Security Agenda (GHSA) advocates for UHC policies that enhance primary health care access, particularly in low- and middle-income countries (LMICs) (Frenk & Moon, 2013).

2. *Workforce Wellness and Well-Being and Ethical Labor Practices.*
A sustainable healthcare delivery system prioritizes employee wellness and well-being, fair wages, and safe working conditions. The National Academy of Medicine (NAM) recommends burn-out prevention strategies, such as mental health support programs and flexible work arrangements, to enhance provider well-being, recruitment, and retention (Shanafelt et al., 2017). Policies that ensure gender equity, diversity in healthcare leadership, and protection against workplace discrimination further contribute to social sustainability in healthcare operations (OECD, 2020).

3. *Community Engagement and Health Literacy.*
Healthcare policies incorporating the principles of socially-responsible sustainability practices emphasize community-based models of care, patient education programs, and culturally competent healthcare services delivery. Expanding mobile health clinics, telemedicine services, and digital health literacy campaigns enhances healthcare accessibility for underserved, vulnerable, and marginalized populations (Fiscella & Sanders, 2016). Furthermore, partnerships between governments, NGOs, and private sector organizations strengthen community health initiatives and improve long-term healthcare sustainability (Rottingen et al., 2017).

4.2.3 Sustainable Governance and Ethical Leadership in Global Healthcare Operations and Policy Design and Development.

Governance in sustainability-aligned global healthcare delivery systems focuses on transparency, accountability, and ethical decision-making. Strong regulatory frameworks ensure that healthcare policies and operations adhere to high ethical standards while promoting innovation and patient safety.

1. *Regulatory Compliance and Policy Transparency.*
    Governments and healthcare organizations must adopt clear regulatory frameworks that promote sustainability compliance. Policies such as the EU Green Deal and the U.S. Inflation Reduction Act include provisions for healthcare institutions to disclose their GHG inventory, implement climate resilience strategies, and adhere to sustainable healthcare guidelines (OECD, 2020). Transparent reporting mechanisms, such as sustainability disclosure requirements for healthcare institutions, help track progress and promote accountability (WHO, 2021).

2. *Public-Private Partnerships for Sustainable Healthcare Innovation.*
    Public-private partnerships (PPPs) are essential for advancing sustainable healthcare initiatives. Programs such as the Global Fund and Gavi, the Vaccine Alliance, harness public and private sector investments to enhance healthcare access and bolster global health security (Røttingen et al., 2017). Furthermore, policies that promote impact investing and social entrepreneurship in healthcare drive long-term financial sustainability and foster innovation (Porter & Lee, 2013).

3. *Cybersecurity and Data Ethics in Healthcare Governance.*
    As healthcare becomes increasingly digitalized, cybersecurity and data privacy regulations are critical governance components. Policies such as the General Data Protection Regulation (GDPR) and the Health Insurance Portability and Accountability Act (HIPAA) ensure patient data protection and ethical use of digital health technologies (Bodenheimer & Sinsky, 2014). Ethical AI frameworks in healthcare further support bias-free clinical decision-making and patient data security (OECD, 2020).

4.2.4 Conclusion.

Integrating the principles of sustainability's ESG foundational pillars into global healthcare delivery system policies and operations is essential for enhancing stewardship, sustainability, equity, and ethical governance. Environmental stewardship and sustainability efforts focus on reducing GHG emissions, promoting sustainable procurement, and managing healthcare waste responsibly. Social sustainability ensures equitable healthcare access, workforce well-being, and patient-centered care models. Strong governance structures, including policy transparency, regulatory compliance, and cybersecurity measures, further strengthen healthcare delivery systems worldwide. By embedding principles of sustainability into healthcare policies and operations, global healthcare delivery systems can achieve long-term sustainability, resilience, and improved patient outcomes.

4.3 The Financial Benefits of Aligning Global Healthcare Systems with U.N. Global Sustainability Development Goals.

Aligning global healthcare delivery systems with the United Nations' Sustainable Development Goals (SDGs) presents significant financial benefits for healthcare organizations, governments, and society. The SDGs, particularly Goal 3 (Good Health and Well-Being), Goal 7 (Affordable and Clean Energy), and Goal 13 (Climate Action), emphasize sustainable healthcare services delivery that reduce costs, improve efficiency, and enhance long-term financial sustainability (United Nations, 2015). By integrating sustainability into healthcare policies and operations, healthcare delivery systems can achieve cost savings, resource optimization, and increased economic resilience while improving population health outcomes.

4.3.1 Cost Savings Through Global Sustainable Healthcare Infrastructure.

Healthcare facilities around-the-world are among the most energy-intensive institutions, consuming large amounts of electricity, water, and resources. Transitioning to renewable energy sources, such as solar and wind, and implementing energy-efficient hospital designs can significantly reduce operational costs. Studies indicate that hospitals adopting green energy solutions reduce energy costs by 20–30% annually (Karliner et al., 2020). The UK's National Health Service (NHS) Greener Plan has already saved over $120 million annually through sustainability-focused energy efficiency initiatives (NHS, 2021).

Effective waste reduction and recycling programs lower hospital expenses associated with waste disposal. The World Health Organization (WHO) estimates that hospitals can reduce waste management costs by 40% by implementing waste segregation, reusable medical supplies, and circular economy practices (WHO, 2020). Additionally, innovations in biodegradable medical equipment and sustainable packaging further contribute to financial savings while promoting environmental sustainability.

4.3.2 Lower Healthcare Costs Through Preventive and Population Health Strategies.

Aligning healthcare with the SDGs involves shifting focus from hospital-centric reactive care to proactive and community-based integrated systems of health. Health prevention—including vaccination programs, early disease detection, and health education—reduces the financial burden of treating chronic diseases. Studies show that every $1 invested in preventive healthcare services saves $4 in long-term medical costs by reducing avoidable hospital admissions and emergency room visits (Maciosek et al., 2017).

SDG-aligned healthcare policies emphasize digital health innovations such as telemedicine, mobile health applications, and

AI-powered diagnostics. The World Economic Forum (WEF) reports that telemedicine reduces healthcare costs by 30–40% while expanding access to care, particularly in rural and underserved regions (WEF, 2022). Countries implementing telehealth solutions have observed millions in annual cost savings through reduced hospital congestion and improved chronic disease management (OECD, 2021).

4.3.3 Economic Growth and Job Creation in Sustainable Global Healthcare Delivery.

Investments in sustainable healthcare infrastructure design, such as eco-friendly hospitals and net-zero buildings, generate significant economic returns while creating new jobs. The International Labour Organization (ILO) estimates that transitioning to a green healthcare economy could create 18 million new jobs globally by 2030 (ILO, 2019). Green hospital construction and retrofitting projects boost local economies while reducing long-term healthcare system expenses.

Pharmaceutical companies aligning with achieving SDGs benefit from cost reductions in sustainable drug manufacturing, ethical supply chain management, and reduced environmental compliance fines. A report by Deloitte (2021) found that sustainability-compliant pharmaceutical firms experienced a 15% increase in investor confidence due to improved risk management and sustainability-driven operational efficiency.

4.3.4 Resilience Against Economic and Climate-Related Healthcare Disruptions.

Climate change-related disasters, such as heatwaves, pandemics, and extreme weather events, increase healthcare costs due to rising patient demand and infrastructure damage. Aligning with SDG 13 (Climate Action) through climate-resilient healthcare facilities and disaster preparedness plans minimizes financial losses from climate-related disruptions. The WHO estimates that climate-resilient healthcare systems could save $2–4 billion annually by reducing climate-induced health burdens (WHO, 2021).

Sustainable healthcare funding models, including impact investing, social bonds, and public-private partnerships (PPPs), reduce the financial strain on governments supporting healthcare delivery. Programs such as Gavi, the Vaccine Alliance, and the Global Fund leverage private sector investments to fund critical healthcare initiatives, resulting in an estimated $47 billion return on investment (ROI) in global health by 2030 (Rottingen et al., 2017).

4.3.5 Conclusion.

Aligning global healthcare delivery systems with the U.N. Sustainable Development Goals offers significant financial advantages, including reduced operational costs, improved resource efficiency, economic growth, and enhanced resilience against climate-related risks. Sustainable healthcare infrastructure design, proactive models of integrated systems of health investments, digital health technologies, and public-private partnerships drive long-term cost savings and financial sustainability. By embedding SDG principles into healthcare policies and operations, global healthcare delivery systems can achieve both financial and public health benefits, ensuring equitable and sustainable healthcare for future generations.

4.4 References.

1. Allen, J. G., MacNaughton, P., Laurent, J. G. C., Flanigan, S. S., Eitland, E. S., & Spengler, J. D. (2015). Green buildings and health. *Current Environmental Health Reports, 2*(3), 250-258.
2. Basu, S., Landrigan, P. J., & Sheikh, K. (2022). Health, equity, and sustainability in healthcare systems. *The New England Journal of Medicine, 387*(12), 1059-1068.

3. Bhutta, Z. A., Lassi, Z. S., Pariyo, G., & Huicho, L. (2017). Global health leadership: The role of universities in building health systems resilience. *Global Health Action, 10*(1), 1353383.

4. Bilec, M. M., Ries, R. J., & Matthews, H. S. (2019). Sustainable hospital design: Influencing healthcare facility resilience. *Building and Environment, 148*, 81-91.

5. Bodenheimer, T., & Sinsky, C. (2014). From triple to quadruple aim: Care of the patient requires care of the provider. *Annals of Family Medicine, 12*(6), 573-576.

6. Deloitte. (2021). *Sustainable healthcare: ESG and financial performance in the life sciences sector.* Deloitte Insights.

7. Dunn, P., Hazzard, E., & Gibson, C. (2019). Data governance and cybersecurity in healthcare. *Health Affairs, 38*(3), 350-357.

8. Eckelman, M. J., & Sherman, J. (2016). Environmental impacts of the U.S. health care system and effects on public health. *PloS ONE, 11*(6), e0157014.

9. Frenk, J., Gómez-Dantés, O., & Moon, S. (2010). From sovereignty to solidarity: A renewed vision for global health leadership. *The Lancet, 375*(9712), 1964–1968.

10. Fiscella, K., & Sanders, M. R. (2016). Racial and ethnic disparities in the quality of health care. *Annual Review of Public Health, 37*, 375-394.

11. Hofmarcher, M. M., Quentin, W., & Busse, R. (2021). Governance in healthcare: Approaches, challenges, and future directions. *Health Policy, 125*(9), 1145-1153.

12. International Labour Organization (ILO). (2019). *Jobs in a net-zero emissions economy: Implications for healthcare workforce planning.*

13. Karliner, J., Slotterback, S., Boyd, R., et al. (2020). *Health care's climate footprint: A global analysis.* Health Care Without Harm.

14. Kickbusch, I., & Gleicher, D. (2012). *Governance for health in the 21st century.* World Health Organization.

15. Kruk, M. E., Gage, A. D., Arsenault, C., Jordan, K., Leslie, H. H., Roder-DeWan, S., ... & Pate, M. (2018). High-quality

health systems in the Sustainable Development Goals era: Time for a revolution. *The Lancet Global Health, 6*(11), e1196-e1252.

16. Maciosek, M. V., Coffield, A. B., Flottemesch, T. J., et al. (2017). Greater use of preventive services in U.S. health care could save lives at little or no cost. *Health Affairs, 29*(9), 1656-1660

17. McGain, F., & Naylor, C. (2014). Environmental sustainability in hospitals. *Medical Journal of Australia, 201*(7), 393-395.

18. National health Service (NHS). (2021). *Delivering a net-zero National Health Service: Greener NHS strategy report.*

19. Organization for Economic Cooperation and Development (OECD). (2020). *Governance in healthcare: Transparency and accountability.* Organisation for Economic Co-operation and Development.

20. Organization for Economic Cooperation and Development (OECD). (2020). *Sustainable healthcare systems: Challenges and policy recommendations.* Organisation for Economic Co-operation and Development.

21. Organization for Economic Cooperation and Development (OECD). (2021). *Telemedicine and digital health policy framework for sustainable healthcare systems.* Organisation for Economic Co-operation and Development.

22. Pichler, P. P., Jaccard, I. S., Weisz, U., & Weisz, H. (2019). International comparison of health care greenhouse gas footprints. *Environmental Research Letters, 14*(6), 064004.

23. Porter, M. & Lee, T.H. (2013). The Strategy That Will Fix Health Care. *Harvard Bus Review*, 91(12), 50 – 70.

24. Rottingen, J. A., Ottersen, T., Ablo, A., Arhin-Tenkorang, D., Benn, C., Evans, D., & McCoy, D. (2017). Shared responsibilities for health: A coherent global framework for health financing. *The Lancet, 389*(10081), 1373-1385.

25. Swensen S, Pugh M, McMullan C, Kabcenell A. (2013). *High-Impact Leadership: Improve Care, Improve the Health of Populations, and Reduce Costs.* IHI White Paper. Cambridge, MA: Institute for Healthcare Improvement.

26. United Nations (UN). (2015). *Sustainable Development Goals: The 2030 Agenda for Sustainable Development.*
27. Watts, N., Amann, M., Arnell, N., et al. (2021). The 2021 report of the Lancet Countdown on health and climate change: Code red for a healthy future. *The Lancet, 398*(10311), 1619–1662.
28. World Economic Forum (WEF). (2022). *The economic benefits of telemedicine: Reducing costs and improving healthcare access.* World Economic Forum.
29. World Health Organization (WHO). (2020). *Global health waste management policies: Reducing medical waste and promoting circular economy practices.*
30. World Health Organization (WHO). (2021). *Climate change and health: Building climate-resilient healthcare systems worldwide.*
31. World Health Organization (WHO). (2021). *Leadership and governance in global health: Strengthening resilience and equity.* World Health Organization.
32. World Health Organization (WHO). (2021). *Global report on health equity and sustainability.* World Health Organization.

5.0

Integrating NbS into Healthcare Delivery Models: Primary Health Care (PHC) and Integrated Systems of Health (ISH).

5.1 The Role of Nature-Based Solutions (NbS) in Strengthening Global Primary Health Care (PHC) Systems and Services.

Nature-based Solutions (NbS) provide a transformational approach to strengthening global Primary Health Care (PHC) services by utilizing natural ecosystems to enhance Advanced Primary Care (APC) services, Essential Public Health Functions (EPHF), and community-based care. These solutions engage and empower both individuals and communities, foster interdisciplinary collaboration, and improve healthcare system resilience. Additionally, NbS help promote sustainable, proactive healthcare services, including disease prevention, health protection, health promotion, disease surveillance and response, as well as emergency preparedness and resilience.

Defined by the International Union for Conservation of Nature (IUCN) as actions that utilize natural or modified ecosystems to address societal challenges effectively and adaptively, NbS can be integrated into PHC systems to reduce disease burden, mitigate climate change effects, and create healthier living environments for all

(IUCN, 2020). The United Nations Sustainable Development Goals (SDGs), particularly SDG 3 (Good Health and Well-being), SDG 6 (Clean Water and Sanitation), and SDG 13 (Climate Action), emphasize the need for sustainable healthcare-related interventions that ensure equitable access to healthcare services delivery while minimizing environmental degradation (United Nations, 2015). Implementing NbS in global PHC systems can lead to lower healthcare costs, reduced reliance on pharmaceuticals, improved mental health outcomes, and greater resilience of healthcare infrastructure against environmental threats.

5.1.1 NbS for Evidence-based Clinical Medicine and Public Health Practices.

1. *Reducing Vector-Borne and Waterborne Diseases.*
    Vector-borne and waterborne diseases continue to pose significant public health challenges world-wide, particularly in low-income and climate-vulnerable regions. Diseases such as malaria, dengue, and cholera are exacerbated by environmental degradation, inadequate sanitation, and poor water management. NbS approaches, such as wetland restoration, sustainable urban drainage systems, and agroforestry, help regulate water cycles, improve biodiversity conservation, and reduce the habitats of disease-carrying vectors. For example, mangrove restoration in Bangladesh has reduced stagnant water pools by 40%, leading to a significant decline in mosquito breeding grounds and cases of malaria and dengue (Bauch et al., 2021). Similarly, community-led reforestation projects in Kenya have contributed to improved air and water quality, reducing cases of respiratory infections and diarrheal diseases (WHO, 2021). The use of nature-based wastewater treatment systems, such as constructed wetlands and biofiltration from prairies and floodplains, has also been shown to remove up to 90% of waterborne pathogens, decreasing cholera outbreaks in vulnerable communities (WHO, 2020).

2. *Urban Green Spaces for Air Quality and Respiratory Health.*
    Air pollution significantly contributes to chronic respiratory diseases, cardiovascular conditions, and premature mortality, with urban populations being especially vulnerable. NbS strategies such as urban forests, pocket prairies, tree planting, pollinator and native vegetation landscaping, and rooftop gardens help sequester greenhouse gas, filter air pollutants (particulate matter (PM) 2.5 micrometers or less and nitrogen dioxide (NO_2)), and regulate temperature, leading to improved respiratory health and reduced hospital admissions. Studies indicate that increasing urban tree coverage by 10% can lower air pollution-related mortality by 8% (Nowak et al., 2018). Cities such as Singapore have integrated green infrastructure, including vertical gardens and biophilic urban design, leading to a 20% decrease in air pollution-related hospital visits (Tan et al., 2020). Additionally, urban NbS reduce the urban heat island effect, which is linked to heat-related illnesses and cardiovascular stress, particularly among vulnerable populations such as the elderly and children.

5.1.2 Mental and Physical Health Benefits of NbS in Global PHC.

1. *Green Prescriptions and Therapeutic Landscapes.*
    An emerging approach in global PHC is the integration of green prescriptions—doctor-recommended nature-based activities, such as outdoor exercise, gardening, and forest therapy—for managing non-communicable diseases (NCDs), stress-related disorders, and mental health conditions. Research shows that regular exposure to nature can reduce stress hormone levels (e.g., cortisol) by 30%, alleviate symptoms of depression, and improve overall mental well-being (Bratman et al., 2019). Countries such as New Zealand and Scotland have implemented national green prescription programs, allowing physicians to prescribe time in nature as part of holistic patient care (Marselle et al., 2021). Japan's "Shinrin-yoku" (forest bathing) program, which involves

immersing oneself in forest environments, has demonstrated significant reductions in anxiety, depression, and blood pressure, providing a low-cost, non-pharmacological intervention for mental and cardiovascular health (Hansen et al., 2017).

2.   *Nature-Based Physical Activity for Health Promotion.*
Chronic diseases such as obesity, type 2 diabetes, respiratory, and cardiovascular disease impose a substantial burden on healthcare systems worldwide. Incorporating nature-based physical activity options—including walking trails, outdoor fitness zones, and nature-integrated urban planning—can encourage active lifestyles and reduce the prevalence of NCDs. Studies have shown that access to green spaces increases physical activity by 50%, leading to lower rates of obesity and related complications (Marselle et al., 2021). In Copenhagen, Denmark, investments in green public spaces have led to a 20% increase in outdoor physical activity, translating to lower healthcare costs and improved public health outcomes (WHO, 2020).

5.1.3 NbS for Strengthening Healthcare Services Delivery Infrastructure and Resilience.

1.   *Climate-Resilient Health Facilities.*
The increasing frequency of climate-related disasters—such as heatwaves, floods, and hurricanes—poses severe threats to healthcare infrastructure and service delivery. NbS solutions, including green roofs, permeable pavements, transitioning to lawns of native grass and forbes, and reforestation, enhance the resilience of healthcare facilities by mitigating extreme heat, reducing flood risks, and improving energy efficiency and fossil fuel reliance. The Philippines' "Green Hospitals" initiative has successfully integrated NbS into hospital design, leading to 25% lower temperatures inside healthcare facilities and reduced electricity consumption (Gonzalez & Marquez, 2021). In addi-

tion, coastal hospitals in the Caribbean have adopted mangrove-based flood defenses, reducing storm-related damages by 30% and ensuring continuity of healthcare services during extreme weather events (WHO, 2021).

2. *Sustainable Water and Food Systems.*
Reliable access to clean water and nutritious food is fundamental for effective PHC. NbS plays a vital role in securing water supplies, improving sanitation, and promoting sustainable subsistence agriculture, thereby reducing the incidence of diarrheal diseases, malnutrition, and waterborne illnesses. In Morocco, the Fog Water Collection project has provided rural PHC centers with a sustainable water supply, reducing waterborne disease rates by 60% and ensuring continuous operation of healthcare services (UNEP, 2021). Similarly, community-led agroecology programs in Africa and Latin America have improved dietary diversity and nutritional health through the promotion of permaculture, organic farming, and indigenous food systems (FAO, 2021). By integrating NbS into food and water security strategies, individual and community PHC services delivery can become more sustainable, cost-effective, and resilient to climate change.

5.1.4 Conclusion.

Nature-based Solutions (NbS) offer a transformational and sustainable approach to strengthening global Primary Health Care (PHC) by enhancing disease prevention, improving mental and physical health, and increasing healthcare infrastructure resilience. By leveraging natural ecosystems for disease control, urban greening, green prescriptions, and climate-resilient healthcare design, NbS provides a cost-effective, environmentally sustainable strategy for addressing pressing global health challenges. As climate change, urbanization, and health inequities continue to strain global healthcare delivery systems, integrating NbS into PHC models will be essential for ensuring equitable, effective, and resilient healthcare for all.

5.2 The Role of Nature-Based Solutions (NbS) in Strengthening Global Integrated Systems of Health (ISH).

As Anthropocene climate change, biodiversity loss, and environmental degradation increasingly threaten human health, integrating NbS into Integrated Systems of Health (ISH) has become a strategic approach to improving resilience and sustainability in healthcare services delivery and public health practice. Previously discussed, NbS, as defined by the International Union for Conservation of Nature (IUCN), refers to "actions to protect, sustainably manage, and restore natural or modified ecosystems that address societal challenges effectively and adaptively" (Cohen-Shacham et al., 2016). When embedded within ISH frameworks, NbS can enhance disease prevention, health protection, disease surveillance and response, mitigate climate-related health risks, implement emergency preparedness and resilience, and promote positive health, optimized wellness, and enhanced well-being through ecosystem-based interventions.

5.2.1 Enhancing Public Health through Ecosystem Services.

Natural ecosystems provide essential ecosystem services, such as clean air, water purification, climate regulation, and food security, which are critical for public health (Bratman et al., 2019). Research has consistently shown that ecosystem degradation leads to increased risks of infectious diseases, respiratory illnesses, and food- and water-borne diseases (Myers et al., 2017).

1. *Air Quality Improvement.*
    Urban tree canopies and forests reduce air pollution by filtering harmful particulate matter (PM2.5), ozone, and nitrogen oxides, leading to lower respiratory and cardiovascular disease

rates (Nowak et al., 2018). For example, a study in the United States estimated that urban trees remove 711,000 metric tons of air pollution annually, preventing thousands of premature deaths (Nowak et al., 2014).

2. *Water Purification and Disease Prevention.*
Wetland restoration is a critical NbS for maintaining water quality. Healthy wetlands filter pollutants, reduce heavy metal contamination, and decrease waterborne disease prevalence, such as cholera and dysentery (Rey et al., 2019). A study in sub-Saharan Africa found that wetland restoration significantly lowered diarrheal disease incidence by improving drinking water sources (Woodward et al., 2020).

3. *Biodiversity and Zoonotic Disease Prevention.*
Deforestation and habitat destruction increase human exposure to zoonotic diseases like Ebola, COVID-19, and Lyme disease (Keesing et al., 2010). Maintaining biodiversity conservation reduces the risk of pathogen spillover by keeping disease reservoirs in check (Rohr et al., 2020).

5.2.2 NbS in Climate Change Adaptation and Mitigation.

Climate change exacerbates health risks such as vector-borne diseases, heat-related illnesses, and food insecurity (Watts et al., 2021). NbS interventions, such as green infrastructure, urban tree canopies, and wetland conservation, help mitigate these risks by lowering urban heat island effects, regulating temperatures, and improving food resilience.

1. *Heat Mitigation through Green Infrastructure.*
Urban heat islands (UHIs) increase morbidity and mortality, particularly among vulnerable populations. Green roofs, urban forests, and tree-lined streets lower ambient temperatures by providing shade and cooling through evapotranspiration (Ziter

et al., 2019). Research shows that increasing urban green spaces by 10% can reduce mortality from extreme heat by up to 40% in some cities (Schwarz et al., 2015).

2. *Vector-Borne Disease Control.*
   Climate change expands the geographic range of vector-borne diseases, including malaria, dengue, and Lyme disease (Rocklöv & Dubrow, 2020). NbS strategies, such as wetland restoration and conservation of mosquito predators, have proven effective in controlling disease vectors. For example, mangrove forests reduce mosquito populations by providing habitat for larval predators, leading to lower malaria transmission (Murdock et al., 2017).

3. *Food Security and Agricultural Resilience.*
   Agroforestry, permaculture, and regenerative agriculture improve soil health, increase crop yields, and enhance climate resilience. Studies in sub-Saharan Africa have demonstrated that agroforestry increases food security while reducing land degradation (Lasco et al., 2014).

5.2.3 Biodiversity Conservation and One Health Approaches.

NbS aligns with One Health principles, which emphasize the inextricable interconnectedness of human, animal, and environmental health.

1. *Preventing Disease Spillover.*
   Zoonotic diseases account for 75% of emerging infectious diseases world-wide (Jones et al., 2008). Deforestation, agricultural expansion, and illegal wildlife trade facilitate cross-species transmission. Restoring natural habitats and reducing human-wildlife contact can help prevent communicable disease outbreaks (Cunningham et al., 2021).

2. *Antimicrobial Resistance and Sustainable Agriculture.*
    Excessive antibiotic use in livestock contributes to antimicrobial resistance (AMR), a growing global health threat. NbS approaches, such as rotational grazing, organic farming, and habitat restoration, reduce antibiotic reliance while maintaining livestock health (Van Boeckel et al., 2017).

3. *Wildlife Conservation for Disease Prevention.*
    Preserving biodiversity through conservation strategies buffers against disease transmission by diluting pathogen reservoirs (Keesing et al., 2010). For example, studies have shown that higher mammalian biodiversity reduces the prevalence of Lyme disease in North America (Ostfeld & Keesing, 2012).

5.2.4 NbS in Mental Health, Wellness, and Well-being.

Exposure to nature has well-documented benefits for mental health, reducing stress, anxiety, and depression (White et al., 2019).

1. *Nature Therapy and Hospital Healing Gardens.*
    Hospital gardens and nature-based therapy improve patient recovery times, reduce stress, and enhance overall patient well-being. A study found that patients recovering from surgery who had a view of nature required fewer painkillers and had shorter hospital stays (Ulrich, 1984).

2. *Green Spaces and Psychological Health.*
    Regular access to parks, forests, and blue spaces (rivers, lakes, and oceans) is linked to lower rates of psychiatric disorders, improved cognitive function, and increased social cohesion (Kondo et al., 2018). A meta-analysis found that people who spent at least 120 minutes in nature per week reported significantly better health and well-being (White et al., 2019).

5.2.5 Policy and Governance Considerations for NbS in ISH.

For NbS to be successfully integrated into ISH, supportive policies, cross-sector collaboration, and sustainable financing mechanisms are essential.

1. *Global Policy Frameworks Supporting NbS.*
   WHO and UNEP advocate for NbS in healthcare design and planning to improve resilience to environmental health risks (WHO, 2020). The European Green Deal promotes NbS as a core strategy for sustainable healthcare delivery systems (European Commission, 2021). The Convention on Biological Diversity (CBD) emphasizes ecosystem conservation for global health benefits (CBD, 2020).

2. *Health-related Financial Investments in NbS.*
   Innovative financing, such as green bonds and public-private partnerships, can help scale up NbS interventions in healthcare and urban planning (Cohen-Shacham et al., 2016).

5.2.6 Conclusion.

Nature-Based Solutions (NbS) play a vital role in strengthening Integrated Systems of Health (ISH) by promoting ecosystem resilience, reducing environmental health risks, and enhancing population wellness and well-being. As global health challenges intensify over the next decade or two, investing in NbS offers a sustainable, cost-effective strategy to safeguard human and planetary health. Future research should focus on quantifying the long-term health benefits of NbS and scaling up implementation across diverse healthcare and public health systems.

5.3 Case studies: Successful Implementation of Nature-Based Solutions (NbS) in Diverse Healthcare Settings.

Nature-based solutions (NbS) integrate natural elements into human environments to enhance health, wellness, well-being, resilience, and sustainability. In healthcare settings, NbS have been utilized to improve patient recovery, reduce stress among healthcare professionals, and promote sustainable healthcare infrastructure. These short case studies examine successful implementations of NbS across three diverse healthcare settings: a metropolitan hospital, a rural community clinic, and a long-term care facility.

Case 1: Urban Hospital – Green Roof and Healing Gardens Location: Bellevue Hospital, New York City, USA.
Intervention: Bellevue Hospital introduced a green roof and healing gardens to provide patients and staff access to nature. The initiative aimed to reduce stress, improve air quality, and support biodiversity in an urban environment.
Outcomes: A study conducted six months post-implementation revealed a 20% reduction in patient-reported stress levels and improved recovery rates for post-operative patients. Additionally, hospital staff reported enhanced job satisfaction and reduced burnout (Ulrich, 2021).

Case 2: Rural Community Clinic – Integrating Agroforestry and Medicinal Gardens Location: Maternal and Child Health Clinic, Kisumu, Kenya.
Intervention: The clinic integrated agroforestry and medicinal gardens to provide herbal remedies, nutritional supplements, and shaded areas for patients. The initiative aimed to enhance local healthcare self-sufficiency and provide culturally relevant healthcare solutions.
Outcomes: The clinic reported a 30% increase in patient adherence to treatment plans, particularly among mothers utilizing herbal supplements for maternal health. Additionally, staff reported a greater

sense of connection to traditional healing practices, improving trust in medical care. (Van den Berg & Custers, 2022).

Case 3: Long-Term Care Facility – Therapeutic Horticulture for Elderly Residents Location: Evergreen Senior Living Center, Ontario, Canada. *Intervention:* A therapeutic horticulture program was introduced to engage elderly residents in gardening activities, including growing vegetables, flowers, and native plants. The goal was to enhance cognitive function, reduce depression, and encourage social interaction. *Outcomes:* Over one-year, cognitive assessments demonstrated a 15% improvement in memory retention among residents participating in gardening activities. Furthermore, depression scores decreased by 25% based on standardized psychological evaluations (Kaplan, 2023).

These three cases highlight the diverse applications of NbS in healthcare settings, from urban hospitals to rural clinics and long-term care facilities. Each intervention demonstrated measurable improvements in patient outcomes, staff well-being, and healthcare sustainability. As global healthcare delivery systems continue to evolve, integrating NbS can play a crucial role in fostering holistic, comprehensive, integrated patient-centered primary health care.

5.4 References.

1. Bauch, S. C., Birkenbach, A. M., Pattanayak, S. K., & Sills, E. O. (2021). Public health impacts of ecosystem change in coastal Bangladesh: A modeling study. *Environmental Research Letters, 16*(3), 035006.
2. Bratman, G. N., Anderson, C. L., Berman, M. G., et al. (2019). Nature and mental health: An ecosystem service perspective. *Science Advances, 5*(7), eaax0903.
3. Cohen-Shacham, E., Walters, G., Janzen, C., & Maginnis, S. (2016). *Nature-based solutions to address global societal challenges.* IUCN.

4. Cunningham, A. A., Daszak, P., & Wood, J. L. N. (2021). One Health, emerging infectious diseases, and wildlife: Two decades of progress? *Philosophical Transactions of the Royal Society B: Biological Sciences, 376*(1837), 20200352.

5. Dasgupta, S., Laplante, B., Meisner, C., Wheeler, D., & Yan, J. (2020). The impact of sea-level rise on developing countries: A comparative analysis. *Climatic Change, 99*(3-4), 389-407.

6. European Commission. (2021). *The European Green Deal.*

7. Gonzalez, J., & Marquez, M. (2021). *Sustainable healthcare infrastructure: Climate resilience in the Philippines' Green Hospitals initiative.* WHO.

8. Hansen, M. M., Jones, R., & Tocchini, K. (2017). Shinrin-yoku (forest bathing) and nature therapy: A state-of-the-art review. *International Journal of Environmental Research and Public Health, 14*(8), 851.

9. IUCN. (2020). *Nature-based solutions to address global societal challenges.* International Union for Conservation of Nature.

10. Kaplan, S. (2023). "Therapeutic Horticulture and Mental Health: A Longitudinal Study in Elderly Care Settings." Aging & Mental Health, 28(1), 55-72.

11. Kondo, M. C., Fluehr, J. M., McKeon, T., & Branas, C. C. (2018). Urban green space and its impact on human health. *International Journal of Environmental Research and Public Health, 15*(3), 445.

12. Lasco, R. D., Delfino, R. J. P., & Espaldon, M. L. O. (2014). Agroforestry systems: A climate change adaptation and mitigation option for smallholder farmers. *Agriculture, Ecosystems & Environment, 187*, 37-46.

13. Laxminarayan, R., Van Boeckel, T., Frost, I., Kariuki, S., & Holloway, K. (2020). The epidemiology of antimicrobial resistance and One Health implications. *Nature Reviews Microbiology, 18*(7), 435-448.

14. Marselle, M. R., Stadler, J., Korn, H., Irvine, K. N., & Bonn, A. (2021). Biodiversity and health in the urban environment. *Science of the Total Environment, 759*, 143612.

15. Murdock, C. C., Sternberg, E. D., & Thomas, M. B. (2017). Malaria transmission and the urban environment. *PloS Neglected Tropical Diseases, 11*(1), e0005515.

16. Nowak, D. J., Hirabayashi, S., Bodine, A., & Greenfield, E. (2018). Tree and forest effects on air quality and human health in the United States. *Environmental Pollution, 242*(B), 1567-1577.

17. Ostfeld, R. S., & Keesing, F. (2012). Effects of host diversity on infectious disease. *Annual Review of Ecology, Evolution, and Systematics, 43*, 157-182.

18. Tan, P. Y., Zhang, J., & Ang, L. (2020). Assessing the ecosystem services of urban greenery: The case of Singapore. *Urban Forestry & Urban Greening, 54*, 126774.

19. Twohig-Bennett, C., & Jones, A. (2018). The health benefits of the great outdoors: A systematic review and meta-analysis of greenspace exposure and health outcomes. *Environmental Research, 166*, 628-637.

20. Ulrich, R. S. (2021). "The Role of Nature in the Healing Process: Evidence-Based Healthcare Design." Journal of Environmental Psychology, 45(3), 245-258.

21. United Nations Environment Programme (UNEP). (2021). *Fog water collection for sustainable drinking water solutions in Morocco.* United Nations Environment Programme.

22. United Nations. (2015). *Sustainable Development Goals: The 2030 Agenda for Sustainable Development.*

23. Van den Berg, A. E., & Custers, M. H. (2022). "Nature-Based Interventions in Global Health: Bridging Traditional and Modern Medicine." International Journal of Public Health, 67(4), 102-115.

24. Watts, N., Amann, M., Arnell, N., Ayeb-Karlsson, S., Belesova, K., Boykoff, M., ... & Costello, A. (2021). The 2021 report of

the Lancet Countdown on health and climate change: Code red for a healthy future. *The Lancet, 398*(10311), 1619-1662.

25. White, M. P., Alcock, I., Grellier, J., et al. (2019). Spending at least 120 minutes a week in nature is associated with good health and well-being. *Scientific Reports, 9*(1), 7730.

26. World Health Organization (WHO). (2020). *WHO Manifesto for a healthy recovery from COVID-19: Prescriptions for a healthy and green recovery.* Geneva: WHO.

27. World Health Organization (WHO). (2020). *Climate resilience and environmentally sustainable health care facilities: Enhancing global health security.*

28. World Health Organization (WHO). (2021). *Nature-based solutions for health: Reducing disease risks and improving public health outcomes.*

6.0

NbS in Environmental, Health, and Safety (EHS) Programs.

Nature-based Solutions (NbS) are increasingly being integrated into Environmental, Health, and Safety (EHS) programs to enhance workplace wellness and well-being, sustainability, and illness/injury risk mitigation. By leveraging natural elements such as green infrastructure, ecological restoration, and biophilic infrastructure design, healthcare and non-healthcare organizations can improve air and water quality, reduce heat stress, and foster healthier environments for employees. Implementing NbS in EHS programs not only strengthens regulatory compliance and operational resilience but also supports biodiversity conservation and climate-change adaptation, creating long-term benefits for both human health and the environment.

6.1 NbS in EHS Programs: Historical Context and Current State.

6.1.1 Introduction.

Environmental, Health, and Safety (EHS) programs are structured frameworks aimed at ensuring workplace safety, minimizing adverse environmental impact, and promoting employee health, wellness, well-being, and resilience. Over time, these programs have evolved in response to industrialization, regulatory mandates, and scientific advancements. In recent years, NbS have been recognized as an essential component of EHS programs, leveraging natural ecosystems' services to enhance environmental stewardship, long-term sustainability, and occupational health. This section explores the historical development of EHS programs, their current state, and the necessity of incorporating NbS for improved health outcomes.

6.1.2 Historical Context of EHS Programs.

The concept of EHS programs can be traced back to the Industrial Revolution, when rapid urbanization and factory-based production led to severe occupational hazards and environmental degradation. Early workplace safety regulations emerged in the late 19th and early 20th centuries, driven by labor movements advocating for better and safer working conditions. For example, the U.K.'s Factory Act of 1833 and the U.S. Occupational Safety and Health Act (OSHA) of 1970 were landmark legislations designed to address worker safety and environmental concerns (Gunningham, 2017). These regulations laid the foundation for contemporary EHS programs by establishing transparent and accountable frameworks and workplace safety standards.

Evolving in parallel with EHS programs, environmental protection efforts gained traction in the mid-20th century with events such as the publication of *Silent Spring* by Rachel Carson in 1962, which highlighted the dangers of chemical pollution. This led to the establishment of regulatory bodies such as the U.S. Environmental Protection Agency (EPA) in 1970 and the implementation of comprehensive environmental laws, including the Clean Air Act and Clean Water Act (Nash, 2000). Over time, EHS programs have expanded beyond regulatory compliance to embrace proactive strategies for sustainability in environmental stewardship, corporate social responsibility, and program governance.

6.1.3 Current State of EHS Programs.

Modern EHS programs encompass a broad range of initiatives, including workplace hazard prevention, emissions control, waste management, and employee wellness programs. Many organizations now adopt international standards such as ISO 14001 (Environmental Management Systems) and ISO 45001 (Occupational Health and Safety), which emphasize continuous improvement and risk management (Grote, 2019). Additionally, advances in digital technologies, including real-time monitoring, data analytics, and AI-driven predictive safety measures, have further strengthened EHS efforts (Bennett et al., 2021).

Despite advancements, EHS programs face growing challenges due to anthropogenic climate change, biodiversity loss, and emerging occupational health risks. Traditional approaches, which rely heavily on engineered solutions, are often costly and resource-intensive. In response, companies and policymakers are increasingly turning to NbS as a value-driven sustainable and cost-effective means to address environmental and health challenges.

6.1.4 Why NbS is Needed in EHS Programs.

NbS enhances EHS programs by integrating ecosystem services to improve air and water quality, mitigate climate risks, and promote employee wellness. Key benefits include:

1. *Improved Air Quality and Health.*
    Green infrastructure, such as urban forests and green roofs, reduces air pollution, thereby decreasing respiratory diseases among employees (Nowak et al., 2018). Plants absorb airborne pollutants and filter particulate matter, which significantly improves indoor and outdoor air quality (Wolch et al., 2014).

2. *Heat Stress Mitigation.*
   Natural shading and vegetation help lower workplace temperatures, reducing heat-related illnesses, particularly in outdoor and industrial settings (Bowler et al., 2010). Studies have shown that urban green spaces can lower surrounding temperatures by up to 5°C, improving comfort and reducing energy costs (Loughner et al., 2012).

3. *Biodiversity Conservation and Ecosystem Resilience.*
   NbS fosters habitat restoration, supporting biodiversity conservation and enhancing ecosystem services that contribute to a healthier environment (Seddon et al., 2020). Workplace landscapes that integrate diverse plant species can attract pollinators and enhance ecological balance (Aronson et al., 2016).

4. *Sustainable Water Management.*
   Rain gardens and wetlands help control stormwater runoff and prevent water contamination, ensuring compliance with environmental regulations (Benedict & McMahon, 2006). Green stormwater infrastructure improves water retention and reduces strain on municipal drainage systems (Fletcher et al., 2015).

5. *Employee Wellness and Productivity.*
   Exposure to natural environments has been shown to reduce stress and improve cognitive function, leading to higher workplace satisfaction and performance (Kaplan & Kaplan, 1989). Studies indicate that employees working in environments with biophilic elements experience a 15% increase in overall productivity (Browning et al., 2014).

6.1.5 Conclusion.

The historical evolution of EHS programs highlights their role in safeguarding worker health and the environment. However, as new climate change challenges emerge, traditional methods alone are insufficient. The integration of NbS into EHS programs offers a holistic, sustainable approach that benefits both businesses and ecosystems. By leveraging natural ecosystems, healthcare and non-healthcare organizations can enhance resilience, reduce operational costs, and contribute to a healthier planet.

6.2 Comparison of EHS and Sustainability Programs.

6.2.1 Introduction.

Environmental, Health, and Safety (EHS) and Sustainability programs are critical frameworks used by both healthcare and non-healthcare organizations to manage health and safety risks, ensure compliance, and promote sustainable business practices. While both share common environmental concerns, their scope, objectives, and implementation differ significantly. This analysis examines the similarities and differences between EHS and Sustainability programs, highlighting how EHS initiatives can align with Sustainability principles to strengthen corporate and stakeholder engagement in addressing Anthropocene climate change.

6.2.2 Comparison of EHS and Sustainability Programs.

Aspect	EHS Programs	Sustainability Programs
Scope	Focuses on regulatory compliance, workplace safety, and environmental impact reduction within operations.	Broader in scope, encompassing environmental stewardship, social responsibility, and corporate governance.

Key Areas	Occupational health and safety, environmental protection, risk management, and regulatory compliance.	Climate change mitigation, diversity & inclusion, ethical business practices, corporate governance, and sustainability reporting.
Primary Drivers	Legal and regulatory requirements (e.g., OSHA, EPA, ISO 14001, ISO 45001).	Investor expectations, shareholder demands, and long-term corporate sustainability goals.
Stakeholders	Employees, regulators, local communities.	Investors, shareholders, customers, employees, regulators, and the broader public.
Implementation	Internal safety policies, environmental risk assessments, compliance audits, and hazard prevention.	Sustainability reporting, greenhouse gas reduction, ethical supply chain management, and governance frameworks.
Measurement & Reporting	Focuses on workplace safety metrics (e.g., injury rates), environmental compliance, and operational safety indicators.	Uses standardized reporting frameworks like GRI, SASB, and TCFD to disclose ESG performance to stakeholders.

On a final note, EHS programs originated primarily from occupational safety and environmental regulations established in the 20th century, including OSHA and EPA guidelines (Gunningham, 2017). Sustainability, on the other hand, emerged as an investment-focused framework in the 21st century, influenced by increasing corporate accountability, investor activism, and global sustainability initiatives such as the United Nations' Sustainable Development Goals (SDGs) (Eccles & Klimenko, 2019).

6.2.3 How EHS Programs Align with the Sustainability Framework.

Although EHS and Sustainability programs have distinct objectives, EHS serves as a foundational element within the sustainability framework. The best alignment strategies include:

1. *Enhancing Environmental Sustainability.*
    EHS programs already focus on pollution control, waste reduction, and compliance with environmental regulations. Expanding EHS strategies to include corporate environmental stewardship goals—such as reducing GHG emissions, enhancing biodiversity conservation, and adopting circular economy principles—strengthens sustainability alignment (Bennett et al., 2021). Research indicates that companies with strong EHS policies report lower environmental liabilities and reduced GHG emissions, aligning with investor expectations for sustainable operations (Kolk, 2016).

2. *Workplace Health and Safety as a Social Responsibility.*
    Social responsibility emphasizes employee wellness and well-being and fair labor practices. EHS programs, with their focus on workplace safety, ergonomics, and mental health initiatives, contribute significantly to the social pillar of sustainable practices by promoting employee welfare, inclusivity, and resilience (Grote, 2019). For example, companies that invest in industrial safety programs report higher employee engagement and productivity (Robson et al., 2007). Additionally, organizations adhering to ISO 45001 improve their reputational standing, making them attractive to socially responsible investors (Zwetsloot et al., 2017).

3. *Governance through Risk Management and Compliance.*
    Strong governance is central to Sustainability programs, and EHS programs play a critical role in workplace risk management, ethical compliance, and operational transparency. Implementing

standardized safety protocols, environmental risk disclosures, and ethical workplace policies ensures alignment with sustainability principles (Nash, 2000). A study by Eccles et al. (2014) found that firms with robust governance structures, including EHS risk oversight, exhibit stronger financial performance and stakeholder trust.

4. *Data-Driven Sustainability Reporting.*
EHS programs generate valuable data on emissions, workplace incidents, and compliance metrics. Integrating this data into sustainability reporting frameworks such as the Global Reporting Initiative (GRI) and Task Force on Climate-related Financial Disclosures (TCFD) enhances transparency, truthfulness, accountability, and helps organizations meet investor expectations (Seddon et al., 2020). For instance, sustainability leaders such as Unilever and Patagonia integrate EHS performance indicators into their sustainability reports to track progress toward climate change mitigation/adaptation goals and workplace safety improvements (Bennett et al., 2021).

5. *Leveraging NbS in EHS for Sustainability Goals.*
Nature-based Solutions (NbS) implemented within EHS programs, such as green infrastructure, greenhouse gas sequestration, and ecosystem restoration, directly support sustainability goals (Benedict & McMahon, 2006). Companies that incorporate NbS, such as green roofs, sustainable water management, and reforestation projects, contribute to sustainability-driven environmental and social benefits (Seddon et al., 2020). Research has shown that urban green spaces improve air quality, reduce heat stress, and enhance employee well-being, reinforcing both EHS and sustainability priorities (Wolch et al., 2014).

6.2.4 Conclusion.

While EHS and Sustainability programs serve different purposes, they are deeply interconnected. EHS programs provide the operational backbone for achieving sustainability objectives, particularly in environmental sustainability, employee wellness and well-being, and regulatory compliance. By expanding EHS initiatives to align with sustainability principles, healthcare and non-healthcare organizations can enhance corporate responsibility, improve stakeholder trust, and drive long-term sustainable growth. A strategic approach integrating EHS within sustainability frameworks enables companies to mitigate workplace risks, achieve competitive advantages, and demonstrate corporate citizenship.

6.3 Climate-conscious EHS Programs.

Climate-conscious Environmental, Health, and Safety (EHS) programs integrate sustainability principles with traditional EHS functions to address the growing challenges of Anthropocene climate change. These programs go beyond regulatory compliance by incorporating proactive measures such as GHG reduction, energy efficiency, climate risk assessments, and sustainable resource management. By aligning workplace safety, environmental protection, and corporate sustainability goals, climate-conscious EHS programs help both healthcare and non-healthcare organizations mitigate environmental risks, enhance resilience, and contribute to broader sustainability commitments. As global businesses face increasing pressure from regulators, investors, and communities to adopt sustainable practices, integrating climate-focused strategies within EHS frameworks is essential for long-term operational success and environmental stewardship.

6.3.1 Occupational Health and Safety within the Context of Nature-based Solutions (NbS).

Occupational Health and Safety (OHS) is a fundamental aspect of workplace management, ensuring employee wellness, reducing hazards for injury and illness, and promoting a safe work environment. In recent years, the integration of Nature-based Solutions (NbS) into OHS frameworks has gained attention as organizations seek sustainable ways to enhance worker safety, mental health, and environmental resilience. Previously discussed, NbS, as defined by the International Union for Conservation of Nature (IUCN), refers to solutions that harness natural processes to address societal challenges, including climate change, disaster risk reduction, and occupational well-being (Cohen-Shacham et al., 2016).

Incorporating NbS into OHS can enhance workplace safety through green infrastructure, improved air quality, climate adaptation, and mental health benefits. This approach aligns with the broader goals of sustainable occupational environments, helping organizations reduce risks while improving employee productivity and well-being.

6.3.1.1 Integrating NbS into Occupational Health and Safety.

1. *Green Infrastructure for Heat and Air Quality Management.*
   a. Work environments, particularly those in construction, agriculture, and manufacturing, often expose employees to extreme heat and poor air quality, leading to heat stress and respiratory illnesses (Kjellstrom et al., 2016).
   b. NbS interventions, such as urban forests, green roofs, and natural shading, help reduce temperatures in industrial settings, mitigating heat stress-related occupational hazards (Bowler et al., 2010).

    c. Green spaces and vegetation also act as natural air filters, reducing exposure to airborne pollutants that contribute to respiratory diseases among workers (Nowak et al., 2014).

2. *Mental Health and Well-being.*
   a. The presence of green spaces in workplaces has been linked to lower stress levels, improved cognitive function, and enhanced overall well-being (Kaplan & Kaplan, 1989).
   b. Studies show that exposure to nature during work hours improves employee morale, reduces anxiety, and decreases workplace burnout (Bratman et al., 2015).
   c. Incorporating NbS elements such as green walls, biophilic office designs, and access to nature-based relaxation areas contributes to better psychological resilience in high-stress occupations, including healthcare and emergency response (Korpela et al., 2017).

3. *Disaster Risk Reduction and Climate Adaptation.*
   a. Extreme weather events, such as floods and storms, result in natural disasters and pose significant risks to workplace safety, especially in industries dependent on outdoor labor (IPCC, 2022).
   b. NbS strategies such as wetland restoration, mangrove barriers, and floodplain conservation can protect industrial zones from climate-related hazards, reducing risks for employees in vulnerable regions (Temmerman et al., 2013).
   c. Additionally, integrating climate-resilient landscapes around work facilities prevents disruptions to operations while safeguarding employees from environmental disasters (Seddon et al., 2020).

4. *Sustainable Worksite Design for Occupational Safety.*
   a. The use of natural materials and eco-friendly building designs in workplace infrastructure enhances indoor environmental

quality, reducing workplace safety risks associated with synthetic pollutants and poor ventilation (Allen et al., 2016).

b. Incorporating water-sensitive urban design (WSUD) and natural drainage systems reduce workplace flooding and contamination risks in industrial zones (Fletcher et al., 2015).

c. NbS-driven safety measures, such as living barriers and vegetative buffers, can also reduce workplace noise pollution and exposure to hazardous chemicals in manufacturing plants (González-Oreja et al., 2010).

6.3.1.2 Conclusion.

Integrating Nature-based Solutions (NbS) into Occupational Health and Safety (OHS) presents a forward-thinking approach to enhancing worker wellness and well-being while addressing climate and environmental challenges. By leveraging green infrastructure, nature-based mental health strategies, disaster resilience planning, and sustainable worksite design, organizations can create safer, healthier, and more climate-resilient workplaces. As climate risks and environmental concerns grow, NbS offers a sustainable pathway for improving occupational health, reducing workplace hazards, and fostering long-term economic and ecological resilience.

6.3.2 Global Environmental Protection and Nature-Based Solutions (NbS).

6.3.2.1 Introduction.

Environmental protection has been a global priority for decades, with nations, organizations, and communities working to mitigate environmental degradation, climate change, and biodiversity loss. Traditionally, biodiversity conservation and environmental protection efforts have relied on regulatory frameworks, technological innovations, and engineered solutions. However, Nature-Based

Solutions (NbS) have emerged as a sustainable, cost-effective, and resilient approach to addressing environmental challenges while providing social and economic benefits (Cohen-Shacham et al., 2016). NbS leverages natural ecosystems and biodiversity conservation to enhance environmental protection, mitigate climate change, and support human well-being at a global scale (Seddon et al., 2020).

This section explores the role of NbS in global environmental protection, with a focus on climate change mitigation, biodiversity conservation, disaster risk reduction, and sustainable resource management.

6.3.2.2 Climate Change Mitigation and Adaptation.

Anthropocene climate change is a major driver of global environmental degradation, affecting ecosystems, economies, and communities. NbS plays a critical role in both *mitigation* (reducing GHG emissions) and *adaptation* (building resilience to climate impacts).

1. *GHG Sequestration through Forest Conservation and Restoration.*
   a. Forests act as major GHG sinks, absorbing one-third of anthropogenic CO_2 emissions annually (Griscom et al., 2017).
   b. Global afforestation and reforestation initiatives, including the Bonn Challenge—launched by the Government of Germany and the IUCN—aim to restore 350 million hectares of degraded land by 2030. Similarly, the Trillion Trees Initiative – launched by WCS, WWF, and Birdlife International–seeks to grow one trillion trees worldwide by 2030, playing a crucial role in climate change mitigation (Laestadius et al., 2015).
   c. Studies indicate that protecting existing forests, particularly in tropical regions, is one of the most cost-effective climate solutions, with the potential to offset 30% of necessary emissions reductions (Brancalion et al., 2019).

2. *Coastal and Marine NbS for Climate Resilience.*
   a. Mangroves, salt marshes, and seagrass meadows provide natural coastal protection, absorbing storm surges and reducing coastal erosion (Duarte et al., 2013).
   b. Mangrove restoration efforts in countries like Bangladesh, Indonesia, and the Philippines have significantly reduced cyclone damage and flood risks, benefiting both biodiversity and local economies (Das & Crépin, 2013).

3. *Urban Green Infrastructure for Heat and Air Pollution Control.*
   a. Cities worldwide are integrating green roofs, urban forests, and wetlands to mitigate the urban heat island effect and reduce air pollution (Ziter et al., 2019).
   b. Cities such as Singapore and Copenhagen have successfully adopted NbS approaches like green corridors and blue-green infrastructure, improving urban resilience to climate change (Tan et al., 2013).

6.3.2.3 Biodiversity Conservation and Ecosystem Restoration.

Biodiversity loss poses significant risks to global ecological stability, food security, and human health. NbS plays a crucial role in protecting and restoring ecosystems that support biodiversity conservation.

1. *Protected Areas and Community-Led Conservation.*
   a. Globally, over 17% of land and 8% of marine areas are designated as protected areas (UNEP-WCMC, 2021). These regions play a critical role in conserving biodiversity through ecosystem-based management.
   b. NbS approaches, such as Indigenous-led conservation programs in the Amazon and Canada's Boreal Forest Initiative, have proven to be highly effective in sustaining biodiversity conservation while supporting Indigenous livelihoods (Schuster et al., 2019).

2. *Restoration of Degraded Landscapes.*
   a. The Great Green Wall Initiative in Africa aims to restore 100 million hectares of degraded land by 2030, combatting desertification while enhancing food security and water access (Sartori et al., 2020).
   b. In China, the Loess Plateau restoration project, which transformed a severely degraded landscape into productive farmland, highlights how NbS can enhance soil health, biodiversity conservation, and local economies (Chen et al., 2019).

3. *Agroecology and Sustainable Land Management.*
   a. Sustainable agricultural practices, such as agroforestry and regenerative farming, reduce land degradation while enhancing biodiversity conservation and food production (Mbow et al., 2019).
   b. The widespread adoption of NbS in agriculture could contribute to over 30% of required greenhouse gas reductions while ensuring food security (Smith et al., 2020).

6.3.2.4 Disaster Risk Reduction and Water Management.

Natural disasters, such as floods, droughts, and wildfires, are exacerbated by climate change and poor environmental management. NbS offers effective, nature-based approaches to disaster risk reduction.

1. *Floodplain Restoration and Wetlands Management.*
   a. Wetlands act as natural buffers against flooding, reducing damage from extreme rainfall events and rising sea levels (Reed et al., 2017).
   b. Countries like the Netherlands and Germany have adopted "Room for the River" programs, restoring natural floodplains to prevent urban flooding (Frijns et al., 2013).

2. *Forest-Based Fire Prevention Strategies.*
   a. Natural forest management and controlled burns, a traditional Indigenous practice, have been reintroduced to reduce wildfire intensity, particularly in fire-prone regions such as Australia, the U.S., and Canada (North et al., 2015).
   b. Reforestation with fire-resistant tree species has been successfully implemented in the Mediterranean and California to reduce wildfire risks (Fernandes et al., 2013).

3. *Water Security and Sustainable Watershed Management.*
   a. NbS-based watershed management programs, such as China's Sponge Cities initiative, use permeable surfaces, urban wetlands, and reforestation to improve urban water security and reduce flood risks (Zhan & Yu, 2020).
   b. Nature-based desalination and wetland conservation in Middle Eastern and North African regions are improving water access in arid environments (González et al., 2018).

6.3.2.5 Sustainable Resource Management and Economic Benefits.

Integrating NbS into resource management ensures long-term ecological health while providing economic benefits.

1. *Sustainable Fisheries and Marine Protected Areas.*
   a. Overfishing and marine degradation have led to declining fish stocks, threatening food security and local economies.
   b. NbS strategies such as no-take marine reserves and community-managed fisheries in the Pacific Islands and Latin America have proven effective in restoring fish populations and increasing long-term yields (Cinner et al., 2016).

2. *Circular Economy and NbS in Waste Management.*
   a. NbS solutions like biodegradable waste recycling, phytoremediation, and nature-based wastewater treatment are

increasingly being used to reduce pollution and recover valuable resources (Keesstra et al., 2018).

  b. Countries such as Sweden and Japan have successfully incorporated NbS principles into zero-waste policies, promoting sustainable resource use (Pires et al., 2019).

6.3.2.6 Conclusion.

Nature-based Solutions (NbS) offer a holistic, cost-effective, and sustainable approach to worldwide environmental risk mitigation and hazard protection. By leveraging ecosystem services, NbS contributes to climate change adaptation and mitigation, biodiversity conservation, disaster risk reduction, and sustainable resource management. Countries across the globe are increasingly integrating NbS into public policy frameworks, urban planning, and corporate sustainability strategies to enhance environmental resilience and economic stability. Expanding the adoption of NbS will be critical in achieving global environmental goals such as the UN Sustainable Development Goals (SDGs) and the Paris Agreement.

6.3.3 Evidence-Based Risk Management in Environmental Health and Safety (EHS) and Nature-based Solutions (NbS).

6.3.3.1 Introduction.

Environmental Health and Safety (EHS) is a multidisciplinary field focused on mitigating risks that impact human health, workplace safety, and the environment. In recent years, the concept of Nature-based Solutions (NbS) has gained prominence as a sustainable approach to addressing environmental challenges such as climate change, biodiversity loss, and disaster risk reduction (Cohen-Shacham et al., 2016). Integrating evidence-based risk management (EBRM) into EHS frameworks within the NbS model is critical for ensuring that nature-driven interventions are both effective and sustainable.

6.3.3.2 Evidence-Based Risk Management in EHS.

Evidence-based risk management (EBRM) within EHS follows a highly structured approach involving hazard identification, risk assessment, control measures, and continuous monitoring (ISO 31000, 2018). EBRM emphasizes the use of empirical data, scientific research, and global best practices to enhance decision-making towards action. (Aven, 2016). Within the NbS framework, this means evaluating the effectiveness of ecosystem-driven solutions through rigorous assessment methodologies.

6.3.3.3 The Role of NbS in Risk Reduction.

NbS refers to strategies that leverage natural processes to address societal and environmental challenges (IUCN, 2020). Examples include urban reforestation for flood mitigation, green infrastructure to improve urban air quality, and wetland restoration for water purification. These solutions align with EBRM objectives of EHS by reducing exposure to all hazards and enhancing resilience. Evidence suggests that NbS can be cost-effective while delivering co-benefits such as biodiversity conservation and GHG sequestration (Kabisch et al., 2017).

6.3.3.4 Green Infrastructure and Flood Risk Management.

One widely studied EBRM engineering solution using the NbS approach is the implementation of green infrastructure for flood mitigation. Research indicates that urban forests, wetlands, and permeable surfaces can significantly reduce flood risk by absorbing excess rainfall and improving drainage capacity (Raymond et al., 2017). Compared to traditional engineering solutions such as concrete flood barriers, NbS approaches are often more adaptable and provide additional environmental benefits.

6.3.3.5 Occupational Health Considerations in NbS.

Implementing NbS within industrial and urban settings requires careful consideration of occupational health risks. For instance, workers involved in large-scale reforestation projects may be exposed to physical hazards such as extreme weather conditions and vector-borne diseases (Seddon et al., 2020). An EBRM approach ensures that mitigation strategies—such as protective equipment, worker training, and exposure assessments—are integrated into NbS initiatives.

6.3.3.6 Challenges and Future Directions in EHS Risk Management.

Despite the advantages of NbS, challenges remain in integrating EBRM solutions into conventional EHS frameworks. These include:

1. *Data Gaps*: Limited long-term studies on the effectiveness of NbS hinder EBRM decision-making (Chausson et al., 2020).
2. *Regulatory Barriers*: Existing EHS regulations often prioritize traditional engineered solutions over EBRM-engineered solutions using NbS approaches (Nelson et al., 2020).
3. *Cost-Benefit Uncertainty*: While EBRM-engineered NbS can offer cost savings over time, initial investments and economic valuations remain complex (Ruiz-Jaen & Aide, 2005).

To overcome these challenges, EHS professionals must collaborate with restoration ecologists, urban planners, and policymakers to develop empirically-supported evidence-based standardized risk assessment frameworks modified for NbS interventions.

6.3.3.7 Conclusion.

Integrating evidence-based risk management (EBRM) into EHS practices using NbS approaches enhances resilience, sustainability, and positive health outcomes. By leveraging scientific research and empirical data, EHS professionals can ensure that nature-driven solutions are both effective and aligned with occupational and environmental risk mitigation strategies. Future advancements should focus on improving data collection, refining regulatory frameworks, and fostering interdisciplinary collaboration to scale up the adoption of NbS in EBRM.

6.3.4 Global Regulatory Guidance in EHS and NbS.

6.3.4.1 Introduction.

Environmental Health and Safety (EHS) regulations are essential for mitigating risks associated with environmental sustainability, workplace safety, and public health practices. As Nature-based Solutions (NbS) emerge as key strategies for addressing Anthropocene climate change, biodiversity loss, and disaster resilience, global regulatory frameworks must evolve to integrate these approaches efficiently, effectively, and sustainably (Cohen-Shacham et al., 2016). Regulatory policies at international, national, and local levels play a crucial role in shaping the adoption, implementation, and standardization of NbS within EHS frameworks.

6.3.4.2 International Regulatory Frameworks Supporting NbS in EHS.

1. *United Nations (UN) and Multilateral Agreements.*
   Several UN-led initiatives promote NbS as part of global environmental and safety regulations:
   a. *The Paris Agreement (2015)*: Encourages the integration of NbS into national climate action plans to meet GHG reduction targets (UNFCCC, 2015).
   b. *The Convention on Biological Diversity (CBD):* Promotes ecosystem-based approaches to mitigate biodiversity loss, aligning with EHS objectives (CBD, 2020).

c. *The Sendai Framework for Disaster Risk Reduction (2015-2030):* Advocates for the use of NbS to enhance disaster resilience through ecosystem-based adaptation (UNDRR, 2015).

2. *International Standards supporting NbS in EHS.*
   a. *ISO 14001 (Environmental Management Systems):* Provides guidance on incorporating empirically-supported ecosystem-based approaches within corporate environmental policies (ISO, 2015).
   b. *ISO 31000 (Risk Management – Guidelines):* Supports evidence-based risk management (EBRM) frameworks that include NbS strategies to mitigate environmental hazards (ISO, 2018).
   c. *IUCN Global Standard for NbS:* Introduces criteria for the effective implementation and scaling of NbS within regulatory and corporate frameworks (IUCN, 2020).

6.3.3.3 Regional and National Regulations on NbS in EHS.

1. *European Union (EU).*
   The EU has taken a leading role in integrating NbS into environmental and workplace safety regulations:
   a. *The European Green Deal (2019):* Prioritizes NbS for achieving GHG neutrality and ecosystem restoration (European Commission, 2019).
   b. *EU Biodiversity Strategy for 2030:* Calls for increased NbS implementation in urban planning, air quality management, and flood mitigation (European Commission, 2020).
   c. *Occupational Safety and Health (OSH) Framework Directive (89/391/EEC):* Ensures that NbS interventions account for worker safety and environmental health risks (EU-OSHA, 2021).

2. *United States.*
   a. *The Clean Water Act (CWA):* Encourages wetland conservation and restoration projects as NbS approaches for improving water quality (EPA, 2020).
   b. *The National Environmental Policy Act (NEPA):* Requires environmental impact assessments that increasingly recognize NbS as viable mitigation measures (CEQ, 2020).
   c. *OSHA Standards:* Address occupational hazards associated with implementing NbS, such as exposure to biological agents in green infrastructure projects (OSHA, 2021).

6.3.4.4 Asia-Pacific Regulations on NbS in EHS.

Countries like China, Japan, and Australia have integrated NbS into national climate resilience strategies:

1. *China's Ecological Red Line Policy (2017):* Protects key ecosystems and promotes NbS in urban development (Xu et al., 2019).
2. *Japan's Climate Adaptation Act (2018):* Includes NbS-based flood risk reduction policies (MLIT, 2020).
3. *Australia's National Climate Resilience and Adaptation Strategy (2021):* Recognizes NbS as essential for disaster risk reduction and biodiversity conservation (Australian Government, 2021).

6.3.4.5 Challenges and Future Directions on NbS in EHS.

Despite progress, regulatory challenges remain:

1. *Standardization Gaps:* The lack of uniform global NbS regulatory guidelines hinders large-scale implementation (Seddon et al., 2020).
2. *Compliance and Enforcement:* Ensuring industries adopt NbS within EHS frameworks requires stronger enforcement mechanisms (Nelson et al., 2020).

3. *Economic Barriers:* Uncertainty in long-term financial returns affects investment in NbS under existing regulatory policies (Ruiz-Jaen & Aide, 2005).

To address these challenges, regulatory bodies should prioritize cross-sector collaboration, standardized EBRM methodologies, and enhanced financial incentives for NbS integration within EHS frameworks.

6.3.4.6 Conclusion.

Global regulatory guidance is evolving to incorporate Nature-based Solutions (NbS) within Environmental Health and Safety (EHS) frameworks. International agreements, regional policies, and national regulations are increasingly recognizing NbS as a viable strategy for enhancing sustainability and evidence-based risk management (EBRM). However, further advancements in regulatory harmonization, enforcement, and economic incentives are needed to fully integrate NbS into mainstream EHS practices.

6.3.5 Case Study: Integrating Industrial Hygiene and NbS for Workplace Health and Sustainability.

Abstract: This case study explores the intersection of industrial hygiene and nature-based solutions (NbS) in mitigating workplace hazards while promoting environmental sustainability. The case focuses on a fictitious manufacturing facility, ABC Manufacturing, that successfully integrated NbS to reduce airborne pollutants, manage heat stress, and improve indoor air quality. The implementation of green infrastructure not only enhanced worker health and productivity but also demonstrated the economic and environmental benefits of such an approach.

Introduction: Industrial hygiene focuses on anticipating, recognizing, evaluating, and controlling environmental factors that may impact worker health. Traditional methods rely heavily on engineering controls, administrative policies, and personal protective equipment (PPE). However, integrating NbS—strategies that utilize natural ecosystem services to address environmental challenges—can offer complementary benefits in improving occupational health and safety (OHS).

Background: A midsized electronics manufacturing company, ABC Manufacturing, faced persistent challenges related to air pollution, heat stress, and volatile organic compound (VOC) exposure in their production facility. Workers reported frequent respiratory issues and heat-related illnesses, leading to increased absenteeism and decreased productivity. The company sought an innovative solution that would address these concerns while aligning with its corporate sustainability goals.

Methodology: The company conducted an industrial hygiene assessment, measuring pollutant levels, thermal comfort indices, and worker health complaints. Key interventions included:
1. *Green Walls and Vegetative Barriers:* Planted green walls indoors and vegetative barriers outside to filter pollutants and improve air quality.
2. *Natural Ventilation and Passive Cooling:* Installed rooftop gardens and tree shading around the facility to lower indoor temperatures and reduce reliance on mechanical cooling.
3. *Biophilic Design Elements:* Integrated natural lighting, plant-based partitions, and water features to improve workplace aesthetics and psychological well-being.

Results: Post-implementation data revealed substantial improvements (what could be expected outcomes):

1. A 40% reduction in VOC concentration levels due to increased air purification from vegetation (Cheng et al., 2022).

2. A 30% decrease in reported heat-related illnesses, attributed to passive cooling measures and improved airflow (Santamouris, 2019).
3. Employee satisfaction surveys indicated an increase in perceived well-being, productivity, and overall job satisfaction.

Discussion: This fictitious case study demonstrates that NbS can complement traditional industrial hygiene measures by addressing both physical and psychological hazards. The cost savings from reduced energy consumption, presenteeism, and absenteeism further support the feasibility of NbS in workplace health management. Challenges expected included initial investment costs and maintenance requirements, which were mitigated through phased implementation and employee engagement programs.

Conclusion: Integrating NbS into industrial hygiene strategies provides a sustainable approach to workplace health and safety. The case study using a fictitious manufacturing company, ABC Manufacturing, underscores the importance of interdisciplinary collaboration between environmental scientists, occupational health professionals, and corporate sustainability leaders to achieve holistic improvements in worker health and environmental resilience.

6.4 The Role of Green Infrastructure in EHS Programs.

6.4.1 Introduction.

Green infrastructure (GI) refers to the strategic use of natural and engineered systems to enhance environmental quality, public health practices, and worker safety. Within Environmental, Health, and Safety (EHS) programs, GI plays a crucial role in mitigating pollution, improving climate resilience, and promoting worker and community positive health and well-being. By integrating nature-based solutions into industrial and urban planning, EHS professionals can align sustainability goals with regulatory compliance and risk management strategies.

6.4.2 Environmental Benefits of Green Infrastructure.

One of the primary objectives of EHS programs is to minimize environmental risks associated with industrial and urban activities. GI solutions such as permeable pavements, green roofs, wetlands, and vegetated swales reduce stormwater runoff, filter pollutants, and improve air quality (U.S. EPA, 2023). These systems help industries comply with the Clean Water Act and other environmental regulations by reducing the volume and contamination of wastewater discharges (Fletcher et al., 2015). Moreover, GI supports biodiversity conservation by creating habitats for wildlife, further enhancing ecosystem services.

6.4.3 Health, Wellness, and Well-Being of Workers and Communities.

Exposure to pollution, extreme heat, and poor air quality presents significant health risks to both workers and surrounding communities. Green infrastructure (GI) helps mitigate these risks by reducing urban heat islands, filtering airborne pollutants, and enhancing mental well-being through access to green spaces (Tzoulas et al., 2007). Workplaces incorporating GI—such as biophilic design elements and vegetative buffers—experience lower employee stress levels and greater overall workplace satisfaction (Korpela et al., 2018). Additionally, urban forests and tree canopies play a vital role in reducing heat stress, protecting outdoor workers from heat-related illnesses, and promoting the overall health, wellness, and well-being of communities (Nowak et al., 2014).

6.4.4 Climate Resilience and Disaster Risk Reduction.

With the increasing frequency of extreme weather events resulting in major natural disasters, EHS programs must account for climate resilience in evidence-based risk management (EBRM). GI enhances infrastructure adaptability by absorbing excess rainwater, reducing flood risks, and stabilizing soil to prevent erosion (Gill et al., 2007).

Companies and municipalities that incorporate GI into their resilience strategies benefit from lower insurance costs, reduced infrastructure damage, and improved emergency preparedness (Kabisch et al., 2017). By addressing these risks proactively, EHS programs contribute to long-term sustainability and business continuity.

6.4.5 Compliance and Cost Efficiency.

Regulatory agencies, including the U.S. Occupational Safety and Health Administration (OSHA) and the Environmental Protection Agency (EPA), increasingly emphasize sustainability in compliance requirements. GI can help organizations meet environmental standards while reducing operational costs. For example, green roofs lower energy costs by providing natural insulation, while constructed wetlands offer cost-effective wastewater treatment solutions (Carter & Keeler, 2008). These approaches not only ensure regulatory adherence but also enhance corporate social responsibility (CSR) initiatives, improving a company's transparency, truthfulness, and trust among its stakeholders.

6.4.6 Conclusion.

Green infrastructure plays a vital role in EHS programs by providing environmental protection, enhancing worker and community health, improving climate resilience, and ensuring regulatory compliance. As organizations continue to integrate sustainability into their evidence-based risk management (EBRM) frameworks, GI offers a practical and cost-effective solution to many of the challenges facing industrial and urban landscapes. Future advancements in green engineering and policy incentives will further drive the adoption of GI as a cornerstone of sustainable EHS strategies.

6.5 Sustainable Resource Management Using NbS in EHS Programs.

6.5.1 Introduction.

Sustainable resource management is a critical component of Environmental, Health, and Safety (EHS) programs, ensuring that industrial and urban activities balance economic growth with environmental preservation and human well-being. One of the most effective approaches to achieving sustainability in EHS is the integration of Nature-based Solutions (NbS)—strategies that leverage natural processes to manage resources efficiently while mitigating environmental risks (IUCN, 2020). By implementing NBS, healthcare and non-healthcare organizations can enhance environmental stewardship, reduce operational costs, and improve compliance with regulatory standards.

6.5.2 Clarifying the Role of Nature-Based Solutions (NbS) in EHS Programs.

Nature-based Solutions (NbS) encompass a broad range of strategies that utilize ecosystems' services and other ecological processes to address environmental and human health challenges. Within the context of EHS programs, NbS includes green infrastructure, natural water filtration systems, reforestation efforts, and circular economy practices to optimize stewardship of resource use while minimizing negative environmental impacts (Cohen-Shacham et al., 2016). These strategies help industries manage risks associated with pollution, climate change, and resource depletion in a way that aligns with long-term sustainability goals.

6.5.3 Water Resource Management Through NBS.

Water scarcity and contamination pose significant risks to industrial operations and surrounding communities. NbS approaches such as constructed wetlands, rainwater harvesting, and bioswales provide cost-effective alternatives to conventional wastewater treatment methods. Constructed wetlands, for instance, use vegetation and microbial processes to remove contaminants from industrial effluents, reducing reliance on chemical treatments and energy-intensive filtration systems (Mitsch & Gosselink, 2015).

By integrating natural water retention measures, industries can also mitigate flood risks and improve groundwater recharge. For example, companies adopting green roofs and permeable pavements can reduce stormwater runoff, lowering pollution loads and meeting regulatory compliance standards such as those outlined in the Clean Water Act (U.S. EPA, 2023).

6.5.4 Sustainable Energy and GHG Management.

Reducing greenhouse gas emissions is a key focus of EHS programs, and NbS offers scalable and sustainable solutions to enhance energy efficiency and GHG sequestration. Urban forests and afforestation projects serve as natural GHG sinks, absorbing atmospheric CO_2 and mitigating the impact of industrial emissions (Griscom et al., 2017).

Moreover, industries are incorporating bioenergy production systems—such as algae-based biofuels and biogas from organic waste—to transition away from fossil fuel dominant energy sources and toward renewable energy alternatives. These approaches contribute to the circular economy by repurposing waste materials into renewable energy, reducing dependency on fossil fuels while promoting sustainable resource cycles (Pauli, 2019).

6.5.5 Circular Economy and Waste Reduction Strategies.

Traditional resource management models in industries often lead to excessive waste production and environmental degradation. The circular economy model, supported by NbS, emphasizes resource efficiency, waste-to-value innovations, and ecosystem-based waste management (Geissdoerfer et al., 2017).

For example, industrial symbiosis networks—where waste from one company becomes a resource for another—are increasingly being adopted in eco-industrial parks. In such settings, organic waste from agricultural or food-processing industries is converted into compost or biogas, reducing landfill dependency while enhancing soil health (Ellen MacArthur Foundation, 2021). Additionally, myco-remediation (the use of fungi to break down pollutants) has emerged as a promising NbS for treating contaminated industrial sites, reducing the need for hazardous chemical remediation processes (Singh, 2017).

6.5.6 Compliance, Cost Savings, and Long-Term Resilience.

Regulatory frameworks worldwide are increasingly promoting nature-based sustainability practices in EHS compliance. The European Union's Green Deal and the United Nations Sustainable Development Goals (SDGs) encourage industries to integrate biodiversity conservation and ecosystem-based adaptation into their resource management strategies (European Commission, 2021).

From an economic standpoint, businesses implementing NbS benefit from lower operational costs, reduced regulatory penalties, and enhanced stakeholder trust. For instance, companies investing in wetland restoration for wastewater treatment report long-term cost savings due to decreased reliance on mechanical treatment systems (Carter & Keeler, 2008). Additionally, nature-inclusive urban planning improves disaster resilience, reducing financial losses linked to climate-related disruptions (Kabisch et al., 2017).

6.5.7 Conclusion.

Sustainable resource management in EHS programs requires innovative, cost-effective, and resilient approaches. Nature-based Solutions (NbS) offer a transformational pathway by integrating ecosystem services and other ecological principles into industrial and urban systems. By enhancing water efficiency, reducing GHG emissions, advancing circular economy practices, and improving occupational health, NbS aligns environmental protection with economic sustainability. As industries and policymakers continue to prioritize green transitions, NbS will remain central to achieving long-term environmental and human health objectives.

6.6 References.

1. Allen, J. G., MacNaughton, P., Satish, U., et al. (2016). Associations of cognitive function scores with greenhouse gas dioxide, ventilation, and volatile organic compound exposures in office workers: A controlled exposure study of green and conventional office environments. *Environmental Health Perspectives, 124(6)*, 805-812.

2. Aronson, M. F., La Sorte, F. A., Nilon, C. H., et al. (2016). A global analysis of the impacts of urbanization on bird and plant diversity. *Biological Conservation, 197*, 272-279.

3. Australian Government. (2021). *National Climate Resilience and Adaptation Strategy 2021–2025.*

4. Aven, T. (2016). Risk assessment and risk management: Review of recent advances on their foundation. *European Journal of Operational Research, 253*(1), 1-13.

5. Benedict, M. A., & McMahon, E. T. (2006). *Green Infrastructure: Linking Landscapes and Communities.* Island Press.

6. Bennett, C. M., McManus, K., & Ramaswamy, S. (2021). Digital transformation of EHS management. *Journal of Safety Research, 79*, 145-159.

7. Bowler, D. E., Buyung-Ali, L. M., Knight, T. M., & Pullin, A. S. (2010). Urban greening to cool towns and cities: A systematic review of the empirical evidence. *Landscape and Urban Planning, 97(3)*, 147-155.

8. Brancalion, P. H. S., Holl, K. D., Strassburg, B. B. N., Rodrigues, R. R., & Gandolfi, S. (2019). Maximizing biodiversity conservation and greenhouse gas stock recovery in tropical forest restoration. *Conservation Letters, 12*(2), e12606.

9. Bratman, G. N., Hamilton, J. P., Hahn, K. S., Daily, G. C., & Gross, J. J. (2015). Nature experience reduces rumination and subgenual prefrontal cortex activation. *Proceedings of the National Academy of Sciences, 112(28)*, 8567-8572.

10. Browning, W. D., Ryan, C. O., & Clancy, J. O. (2014). *14 Patterns of Biophilic Design: Improving Health and Well-Being in the Built Environment.* Terrapin Bright Green.

11. Carter, T., & Keeler, A. (2008). Life-cycle cost-benefit analysis of extensive vegetated roof systems. *Journal of Environmental Management, 87*(3), 350-363.

12. Chausson, A., Turner, B., Seddon, D., et al. (2020). Mapping the effectiveness of nature-based solutions for climate adaptation. *Global Change Biology, 26*(11), 6134-6155.

13. Chen, Y., Wang, K., Lin, Y., Shi, W., Song, Y., & He, X. (2019). Balancing green and grain trade: The case of China's 'Loess Plateau restoration. *Ecological Economics, 159*, 278–292.

14. Cheng, H., Cheung, J., & Wong, M. (2022). The role of urban vegetation in mitigating air pollution and heat stress in industrial settings. *Environmental Research*, 212, 113523.

15. Cinner, J. E., Huchery, C., MacNeil, M. A., et al. (2016). Bright spots among the world's coral reefs. *Nature, 535*(7612), 416–419.

16. Cohen-Shacham, E., Walters, G., Janzen, C., & Maginnis, S. (2016). *Nature-based Solutions to Address Global Societal Challenges.* IUCN.

17. Convention on Biological Diversity, (CBD). (2020). *Global Biodiversity Framework Post-2020.*

18. Council on Environmental Quality, (CEQ). (2020). *National Environmental Policy Act (NEPA) Implementing Regulations.*

19. Das, S., & Crépin, A. S. (2013). Mangroves can provide protection against wind damage during storms. *Estuarine, Coastal and Shelf Science, 134*, 98–107.

20. Duarte, C. M., Losada, I. J., Hendriks, I. E., Mazarrasa, I., & Marbà, N. (2013). The role of coastal plant communities for climate change mitigation and adaptation. *Nature Climate Change, 3*(11), 961–968.

21. Eccles, R. G., Ioannou, I., & Serafeim, G. (2014). The impact of corporate sustainability on organizational processes and performance. *Management Science, 60(11)*, 2835-2857.

22. Eccles, R. G., & Klimenko, S. (2019). The investor revolution. *Harvard Business Review, 97(3)*, 106-116.

23. Ellen MacArthur Foundation. (2021). *Circular economy in industry: Strategies for resource efficiency and sustainability.*

24. Environmental Protection Agency, (EPA). (2020). *The Clean Water Act and Wetland Protection.*

25. European Commission. (2021). *EU Green Deal and nature-based solutions for sustainability.*

26. EU-OSHA. (2021). *EU Occupational Safety and Health (OSH) Framework Directive 89/391/EEC.*

27. European Commission. (2019). *The European Green Deal.*

28. European Commission. (2020). *EU Biodiversity Strategy for 2030.*

29. Fernandes, P. M., Davies, G. M., Ascoli, D., et al. (2013). Prescribed burning in southern Europe: Developing fire management in a dynamic landscape. *Frontiers in Ecology and the Environment, 11*(1), e4–e14.

30. Fletcher, T. D., Shuster, W., Hunt, W. F., Ashley, R., Butler, D., & Viklander, M. (2015). SUDS, LID, BMPs, and more–The evolution and application of terminology surrounding urban drainage. *Urban Water Journal, 12(7)*, 525-542.

31. Frijns, J., Dirkx, J., Van Riet, O., & Van Cleef, R. (2013). Flood risk management strategies in the Netherlands: A policy perspective. *Water Resources Management, 27*(15), 4645–4666.

32. Geissdoerfer, M., Savaget, P., Bocken, N. M., & Hultink, E. J. (2017). The circular economy–A new sustainability paradigm? *Journal of Cleaner Production, 143*, 757-768.

33. Gill, S. E., Handley, J. F., Ennos, A. R., & Pauleit, S. (2007). Adapting cities for climate change: The role of the green infrastructure. *Built Environment, 33*(1), 115-133.

34. González, M. A., León, J. A., & Veiga, F. (2018). Sustainable water management in the Middle East and North Africa through nature-based solutions. *Water Policy, 20*(3), 505–521.

35. González-Oreja, J. A., Bonache-Regidor, C., Buzo-Franco, D., & De La Fuente-Díaz-Ordaz, A. A. (2010). Can human wellbeing and environmental protection be reconciled? Relationships between noise reduction and landscape aesthetics. *Landscape and Urban Planning, 98(1)*, 156-163.

36. Griscom, B. W., Adams, J., Ellis, P. W., et al. (2017). Natural climate solutions. *Proceedings of the National Academy of Sciences, 114*(44), 11645–11650.

37. Grote, G. (2019). Management of uncertainty: Theory and application in the design of safety management systems. *Safety Science, 113*, 435-449.

38. Gunningham, N. (2017). *Regulating Workplace Safety and Environmental Protection*. Routledge.

39. Intergovernmental Panel on Climate Change (IPCC) (2022). *Climate Change 2022: Impacts, Adaptation, and Vulnerability.* Intergovernmental Panel on Climate Change, Sixth Assessment Report.

40. International Organization for Standardization (ISO). (2018). *ISO 31000:2018 - –risk management — Guidelines.*

41. International Union for Conservation of Nature (IUCN). (2020). *Nature-based solutions for societal challenges.* International Union for Conservation of Nature.

42. International Organization for Standardization, (ISO). (2015). *ISO 14001:2015 - –Environmental Management Systems.*

43. International Organization for Standardization, (ISO). (2018). *ISO 31000:2018 - –Risk Management Guidelines.*
44. IUCN. (2020). *Global Standard for Nature-based Solutions.*
45. Kabisch, N., Korn, H., Stadler, J., & Bonn, A. (Eds.). (2017). *Nature-based solutions to climate change adaptation in urban areas.* Springer.
46. Kaplan, R., & Kaplan, S. (1989). *The Experience of Nature: A Psychological Perspective.* Cambridge University Press.
47. Keesstra, S., Geissen, V., Mosse, K., et al. (2018). Soil as a filter for groundwater quality. *Current Opinion in Environmental Science & Health, 5,* 54–63.
48. Kjellstrom, T., Briggs, D., Freyberg, C., Lemke, B., Otto, M., & Hyatt, O. (2016). Heat, human performance, and occupational health: A key issue for the assessment of global climate change impacts. *Annual Review of Public Health, 37,* 97-112.
49. Kolk, A. (2016). The social responsibility of international business: From ethics and the environment to CSR and governance. *Journal of World Business, 51(1),* 23-34.
50. Korpela, K., Borodulin, K., Neuvonen, M., Paronen, O., & Tyrväinen, L. (2017). Analyzing the mediators between nature-based outdoor recreation and emotional well-being. *Journal of Environmental Psychology, 51,* 1-7.
51. Korpela, K. M., De Bloom, J., & Kinnunen, U. (2018). From restorative environments to restoration in work. *Health & Place, 50,* 136-145.
52. Laestadius, L., Maginnis, S., Minnemeyer, S., & Potapov, P. (2015). Bonn challenge: Rationale, implementation, and progress. *World Resources Institute.*
53. Mbow, C., Rosenzweig, C., Barioni, L. G., et al. (2019). Food security. *IPCC Special Report on Climate Change and Land.*
54. Ministry of Land, Infrastructure, Transport and Tourism, (MLIT). (2020). *Japan's National Climate Adaptation Plan.*
55. Mitsch, W. J., & Gosselink, J. G. (2015). *Wetlands.* John Wiley & Sons.

56. Nash, J. (2000). *Regulating Workplace Safety: System and Sanctions.* ILR Press.
57. Nelson, D. R., Adger, W. N., & Brown, K. (2020). Adaptation to environmental change: Contributions of a resilience framework. *Annual Review of Environment and Resources, 32*(1), 395-419.
58. North, M. P., Collins, B. M., & Stephens, S. L. (2015). Using fire to increase the scale, benefits, and future maintenance of fuels treatments. *Journal of Forestry, 113*(5), 421–431.
59. Nowak, D. J., Crane, D. E., & Stevens, J. C. (2014). Air pollution removal by urban trees and shrubs in the United States. *Urban Forestry & Urban Greening, 4(3)*, 115-123.
60. Nowak, D. J., Hirabayashi, S., Bodine, A., & Greenfield, E. (2018). Tree and forest effects on air quality and human health in the United States. *Environmental Pollution, 232*, 311-318.
61. Occupational Safety and Health Administration, (OSHA). (2021). *Occupational Exposure to Biological Hazards in Green Infrastructure Projects.*
62. Pauli, G. (2019). *The blue economy 3.0: The marriage of science, innovation and entrepreneurship creates a new business model that transforms society.* Paradigm Publications.
63. Pires, A., Martinho, G., Ribeiro, R., & Dinis, M. (2019). Waste management practices towards zero waste: A European perspective. *Waste Management, 87*, 459–467.
64. Raymond, C. M., Frantzeskaki, N., Kabisch, N., et al. (2017). A framework for assessing and implementing the co-benefits of nature-based solutions in urban areas. *Environmental Science & Policy, 77*, 15-24.
65. Reed, M. S., Stringer, L. C., Fazey, I., Evely, A. C., & Kruijsen, J. H. J. (2017). Five principles for the practice of knowledge exchange in environmental management. *Journal of Environmental Management, 146*, 337–345.
66. Robson, L. S., et al. (2007). The effectiveness of occupational health and safety management system interventions. *Safety Science, 45(3)*, 329-353.

67. Ruiz-Jaen, M. C., & Aide, T. M. (2005). Restoration success: How is it being measured? *Restoration Ecology, 13*(3), 569-577.

68. Santamouris, M. (2019). Cooling the cities—A review of reflective and green roof mitigation technologies to fight heat island and improve comfort in urban environments. *Solar Energy,* 182, 121-135.

69. Sartori, M., Biber-Freudenberger, L., Epple, C., et al. (2020). The Great Green Wall initiative for the Sahara and the Sahel: Sustainability and feasibility challenges. *Sustainability, 12*(4), 1484.

70. Schuster, R., Germain, R. R., Bennett, J. R., Reo, N. J., & Arcese, P. (2019). Vertebrate biodiversity on Indigenous-managed lands in Australia, Brazil, and Canada equals that in protected areas. *Environmental Science & Policy, 101*, 1–6.

71. Seddon, N., Chausson, A., Berry, P., Girardin, C. A., Smith, A., & Turner, B. (2020). Understanding the value and limits of nature-based solutions to climate change and other global challenges. *Philosophical Transactions of the Royal Society B, 375(1794)*, 20190120.

72. Singh, H. (2017). *Myco-remediation: Fungal bioremediation.* Wiley Blackwell.

73. Smith, P., Adams, J., Beerling, D. J., Beringer, T., Calvin, K. V., Fuss, S., ... & Minx, J. C. (2020). Land-management options for greenhouse gas removal and their impacts on ecosystem services and the Sustainable Development Goals. *Annual Review of Environment and Resources, 45*, 79–107.

74. Tan, P. Y., Wang, J., & Sia, A. (2013). Perspectives on five decades of the urban greening of Singapore. *Cities, 32*, 24–32.

75. Temmerman, S., Meire, P., Bouma, T. J., Herman, P. M., Ysebaert, T., & De Vriend, H. J. (2013). Ecosystem-based coastal defence in the face of global change. *Nature, 504(7478)*, 79-83.

76. Tzoulas, K., Korpela, K., Venn, S., et al. (2007). Promoting ecosystem and human health in urban areas using green infrastructure: A literature review. *Landscape and Urban Planning, 81*(3), 167-178.

77. United Nations Environment Programme World Conservation Monitoring Centre, (UNEP-WCMC). (2021). *Protected Planet Report.*

78. United Nations Office for Disaster Risk Reduction, (UNDRR). (2015). *Sendai Framework for Disaster Risk Reduction 2015-2030.*

79. United Nations Framework Convention on Climate Change, (UNFCCC). (2015). *The Paris Agreement.*

80. United States Environmental Protection Agency (EPA). (2023). *Green infrastructure: Managing water sustainably.*

81. Wright, H., Huq, S., & Reeves, J. (2020). Nature-based solutions for climate resilience in cities. *Climate and Development, 12*(2), 85-97.

82. Wolch, J. R., Byrne, J., & Newell, J. P. (2014). Urban green space, public health, and environmental justice: The challenge of making cities 'just green enough'. *Landscape and Urban Planning, 125*, 234-244.

83. Xu, X., Tan, Y., Yang, G., & Zhang, H. (2019). Ecological Red Line Policy and Urban Development in China. *Environmental Science & Policy, 92*, 1-11.

84. Zhan, J., & Yu, L. (2020). Sponge city construction and ecological security pattern optimization. *Journal of Environmental Management, 264*, 110460.

85. Ziter, C. D., Pedersen, E. J., Kucharik, C. J., & Turner, M. G. (2019). Scale-dependent interactions between tree canopy cover and impervious surfaces reduce daytime urban heat during summer. *Proceedings of the National Academy of Sciences, 116*(15), 7575–7580.

86. Zwetsloot, G. I. J. M., et al. (2017). Corporate social responsibility and safety and health at work. *Safety Science, 94*, 27-45.

7.0

Policy, Advocacy, and Global Leadership in Sustainable Healthcare Delivery Systems.

Sustainable healthcare delivery systems require robust policies, proactive advocacy, and strong global leadership to ensure equitable access to healthcare services delivery while minimizing environmental and economic burdens. Policy frameworks, such as the World Health Organization's (WHO) *Global Strategy on Human Resources for Health* (WHO, 2016) and the United Nations Sustainable Development Goals (SDGs), guide nations in integrating sustainability into healthcare delivery systems. Advocacy efforts, led by healthcare organizations-- the Global Green and Healthy Hospitals Network--drive systemic changes by promoting low-GHG emissions' healthcare models and climate resilience in health infrastructure (Health Care Without Harm, 2021). Global leadership, exemplified by collaborations between the WHO and national governments, fosters cross-border solutions to address challenges such as workforce shortages, climate-related health risks, and digital healthcare transformation (Kickbusch et al., 2019).

By aligning policy, advocacy, and leadership with sustainability goals, global healthcare delivery systems can transition toward resilient, equitable, and resource-efficient models of care, ensuring long-term health security for populations worldwide.

7.1 The Role of Healthcare Leadership in Advocating for NbS Integration into National and Global Healthcare Strategies.

7.1.1 Introduction.

Healthcare leaders play a crucial role in advancing Nature-based Solutions (NbS) as a core component of national and global health strategies. As climate change and environmental degradation increasingly impact the public's health, wellness, and well-being, integrating NbS into healthcare policies and infrastructure design is essential to fostering resilience, sustainability, and equity in global healthcare delivery systems. By leveraging their influence, healthcare leaders can advocate strongly for policy reforms, funding resources, and cross-sector collaborations that promote ecosystem-based healthcare strategies (WHO, 2022).

7.1.2 Policy Advocacy for Nature-based Solutions in Global Healthcare Services Delivery.

Healthcare leaders, including hospital administrators, policymakers, and global health officials, have a unique platform to influence policy development. By advocating for green infrastructure in healthcare facilities, ecosystem-based disease prevention, and sustainable medical supply chains, leaders can help shape policies that integrate NbS into national health strategies (UNEP, 2021). For example, the WHO Manifesto for a Healthy Recovery from COVID-19 highlighted the need for investments in sustainable biophilic infrastructure design, including green spaces and air quality improvements, to mitigate future pandemics (WHO, 2020).

National governments have begun incorporating NbS into health strategies, particularly in climate-vulnerable regions. In the European Union, the EU Green Deal and Biodiversity Strategy encourage the use of NbS to enhance urban health and climate resilience (European Commission, 2021). Healthcare leaders can advocate for similar policies in developing nations, ensuring equitable access to NbS-driven health interventions, such as urban tree canopies to reduce heat-related illnesses (Kabisch et al., 2017).

7.1.3 Promoting Cross-Sector Collaboration and Global Health Leadership.

Healthcare leaders play a vital role in fostering interdisciplinary and multisectoral collaboration between national and local health agencies, academic health institutions, environmentally-focused community-based organizations, and urban planners. International initiatives, such as the Global Green and Healthy Hospitals Network (GGHH), demonstrate how partnerships between hospitals and sustainability experts can lead to significant reductions in healthcare-related GHG emissions and resource consumption (Health Care Without Harm, 2021).

At the global level, institutions like the World Bank and the United Nations Development Programme (UNDP) are investing in NbS for health resilience, recognizing that sustainable land use, water conservation, and pollution control directly impact disease prevention (World Bank, 2022). Healthcare leaders who engage in these global initiatives can bridge the gap between healthy public policies and environmental sustainability, ensuring that NbS is integrated into primary health care planning, infectious disease control, and non-communicable disease prevention.

7.1.4 Integrating NbS into Healthcare Infrastructure and Service Delivery.

One of the most direct ways healthcare leaders can support NbS is through the greening of healthcare facilities. By advocating for energy-efficient hospitals, sustainable procurement policies, and green spaces around medical institutions, leaders contribute to both patient well-being and environmental sustainability (Chivian & Bernstein,

2008). Studies show that patients in hospitals with access to natural environments experience faster recovery times, reduced stress, and improved overall outcomes (Ulrich et al., 2008).

Beyond hospital settings, NbS can be integrated into community-based healthcare strategies, particularly in regions prone to vector-borne diseases, air pollution, and water contamination. Leaders can champion initiatives that promote mangrove restoration for coastal protection against disease outbreaks, reforestation to improve air quality, and wetland conservation for water filtration, aligning health sector strategies with ecosystem restoration efforts (Cohen-Shacham et al., 2016).

7.1.5 Driving Research and Education on NbS in Healthcare.

Healthcare leaders also play a pivotal role in advancing research on NbS and integrating sustainability into all levels of medical education. Academic healthcare institutions and professional healthcare organizations, such as the Lancet Planetary Health Commission, emphasize the need for a current and future healthcare workforce that understands the intersection of environmental and public health practice (Whitmee et al., 2015). By pushing for the inclusion of NbS in workforce development and medical training, leaders ensure that future healthcare professionals are equipped to implement sustainable healthcare interventions.

In research, leaders can support studies that evaluate the cost-effectiveness and long-term benefits of NbS in healthcare settings, strengthening the evidence base for its widespread adoption. Institutions like the United Nations Environment Programme (UNEP) and the Intergovernmental Science-Policy Platform on Biodiversity and Ecosystem Services (IPBES) have identified NbS as a critical tool for achieving sustainable health outcomes, and healthcare leaders can drive policy shifts based on emerging scientific findings (UNEP, 2021).

7.1.6 Overcoming Barriers to NbS Adoption in Health Strategies.

Despite its benefits, the integration of NbS into national and global health strategies faces several challenges, including limited funding, lack of policy coordination, and resistance to change in the health-care sector (WHO, 2022). Healthcare leaders must address these barriers by engaging policymakers and stakeholders to align health and environmental policies, securing funding for NbS projects through partnerships with global organizations and private-sector investors, consulting with climate and environmental scientists on empirically-supported evidence-based NbS to embed in their inter-nal infrastructure for resource sustainability and climate resilience, developing training programs for healthcare professionals to ensure effective implementation of NbS-driven health solutions, and raising public awareness on the role of nature in health protection, health promotion and disease prevention.

7.1.7 Conclusion.

Healthcare leaders play an essential role in advocating for the inte-gration of Nature-based Solutions (NbS) into national and global health strategies. By shaping policy, fostering interdisciplinary col-laboration, promoting sustainable healthcare infrastructure, advanc-ing research, and overcoming systemic barriers, leaders ensure that healthcare delivery systems become more resilient, equitable, and environmentally sustainable. As Anthropocene climate change and environmental degradation continue to pose significant health chal-lenges, the adoption of NbS-driven strategies will be crucial in secur-ing long-term public health and planetary well-being.

7.2 International Climate Change Frameworks Aligning with Global Healthcare Sustainability Mission, Vision, and Goals.

7.2.1 Introduction.

As Anthropocene climate change increasingly threatens the public's health, international climate frameworks have become essential in shaping sustainable healthy public policies worldwide. Aligning global healthcare sustainability with empirically-supported climate change mitigation and adaptation strategies ensures that health systems are resilient, environmentally responsible, and capable of addressing climate-related health burdens. Leading frameworks such as the Paris Agreement (2015), the United Nations Sustainable Development Goals (SDGs), the WHO Operational Framework for Building Climate-Resilient Health Systems (WHO, 2021), and the Glasgow Climate Pact (2021) provide strategic guidance for healthcare sustainability efforts at national and global levels.

7.2.2 The Paris Agreement: Healthcare Decarbonization and Climate Resilience.

The Paris Agreement (2015) is a landmark framework under the United Nations Framework Convention on Climate Change (UNFCCC) that aims to limit global warming to below 2°C, with efforts to keep it at 1.5°C (UNFCCC, 2015). Recognizing the health sector's role in global emissions, the COP26 Health Programme—launched under the Paris Agreement—urges governments to build climate-resilient and low-greenhouse gas healthcare systems (WHO, 2021). This directly aligns with healthcare sustainability goals such as:

1. Reducing GHG emissions of healthcare facilities through energy-efficient hospital infrastructure and sustainable procurement.
2. Developing resilient healthcare systems to withstand climate-induced disasters, including heatwaves, floods, and pandemics.

Countries like the United Kingdom and the United States have committed to net-zero healthcare systems by 2030, integrating the Paris Agreement's climate targets into national health policies (Watts et al., 2021).

7.2.3 The United Nations Sustainable Development Goals (SDGs) and Healthcare Sustainability.

The UN Sustainable Development Goals (SDGs) provide a comprehensive blueprint for aligning climate action with global healthcare sustainability. Key SDGs relevant to healthcare sustainability include:

1. *SDG 3 (Good Health and Well-Being):* Calls for universal healthcare access while addressing climate-related health threats such as air pollution and vector-borne diseases (UNDP, 2019).
2. *SDG 7 (Affordable and Clean Energy):* Advocates for renewable energy adoption in healthcare infrastructure, reducing dependence on fossil fuels (UNEP, 2020).
3. *SDG 13 (Climate Action):* Emphasizes the integration of climate-resilient healthy public policies into national adaptation plans.

These SDGs form the foundation for climate-smart healthcare models, particularly in low- and middle-income countries (LMICs) that suffer disproportionate climate-related health impacts (WHO, 2022).

7.2.4 WHO Operational Framework for Climate-Resilient Health Systems.

The World Health Organization (WHO) Operational Framework for Building Climate-Resilient Health Systems (2021) provides healthcare institutions world-wide with practical strategies to integrate climate change adaptation into their operational and financial policies. The framework outlines multiple critical components including integrated evidence-based risk surveillance and response for climate-sensitive diseases, strengthening healthcare services infrastructure to withstand climate shocks, implementing sustainable medical supply chain management to minimize emissions throughout the entire life cycle of medical products and services. The framework directly supports global healthcare missions that prioritize sustainability, environmental justice, and health equity in the face of climate change.

7.2.5 The Glasgow Climate Pact and Its Impact on Healthcare.

Adopted at COP26 in 2021, the Glasgow Climate Pact reaffirmed global commitments to climate mitigation and adaptation, emphasizing the need to transition towards sustainable, climate-resilient healthcare systems (UNFCCC, 2021). The agreement highlights urgent funding for climate adaptation in healthcare, particularly in vulnerable regions, phasing out fossil fuel subsidies, pushing the health sector toward renewable energy sources (The Lancet, 2021), and strengthening global climate-health partnerships, ensuring that healthcare delivery systems integrate environmental sustainability into national and global health strategies.

7.2.6 The Role of Global Health Institutions in Climate-Health Advocacy.

Several global health institutions are leading initiatives that align climate change frameworks with healthcare sustainability:

1. *Health Care Without Harm (HCWH)* supports net-zero emissions in hospitals, advocating for climate-conscious healthcare procurement (HCWH, 2021).
2. *The Lancet Countdown on Health and Climate Change* tracks global progress on climate-health policies, influencing governments to integrate climate goals into healthcare delivery (Romanello et al., 2021).

3. *The World Bank Climate and Health Initiative* provides financing for climate-resilient health infrastructure in LMICs (World Bank, 2022).

7.2.7 Conclusion.

International climate change frameworks, including the Paris Agreement, SDGs, WHO's Climate-Resilient Health Systems Framework, and the Glasgow Climate Pact, play a pivotal role in aligning global healthcare sustainability goals. By integrating empirically-supported climate action into global healthcare delivery systems, these frameworks help reduce environmental footprints, enhance disaster resilience, and protect the public's health from climate-related risks. Global healthcare institutions, policymakers, and industry leaders must collaborate to accelerate the transition toward sustainable, climate-smart models of health and care that ensure long-term planetary and population health.

7.3 The Need for Interdisciplinary and Multisectoral Collaboration in Adopting NbS.

7.3.1 Introduction.

Interdisciplinary and multisectoral collaboration is essential for integrating Nature-based Solutions (NbS) into public health initiatives and environmental sustainability efforts. As climate change, biodiversity loss, and pollution continue to threaten human health, governments, healthcare institutions, and environmental organizations must work together to implement NbS that strengthen healthcare system resilience, mitigate environmental degradation, and enhance population health, wellness, and well-being (WHO, 2022). Such collaboration ensures that public health policies incorporate ecological considerations while environmental policies acknowledge the direct impact of ecosystems on human health.

7.3.2 The Role of Governments in NbS Implementation.

Governments, at all levels of organization, play a pivotal role in setting policies, funding initiatives, and ensuring regulatory frameworks that facilitate the adoption of NbS in healthcare and environmental management. Several international policies highlight the importance of governmental action in NbS-driven health solutions:

1. The European Green Deal (2019) promotes urban green infrastructure to improve air quality and reduce heat-related illnesses (European Commission, 2021).
2. The United Nations Sustainable Development Goals (SDGs), particularly SDG 3 (Health and Well-Being) and SDG 13 (Climate Action), advocate for integrating NbS into public health strategies (UNEP, 2020).
3. Nationally Determined Contributions (NDCs) under the Paris Agreement (2015) emphasize nature-based climate adaptation strategies, including reforestation and wetland conservation to mitigate vector-borne diseases (UNFCCC, 2015).

Governments must work alongside academic and non-academic healthcare institutions, environmental advocacy groups, and community-based organizations to translate these policies into actionable, locally tailored initiatives that address both public health initiatives and ecosystem sustainability.

7.3.3 Healthcare Institutions as Drivers of NbS Adoption.

Healthcare institutions, including academic medical centers, hospitals, clinics, and public health agencies, play a critical role in integrating NbS into institutional infrastructure, medical supply chain, service delivery, and patient care. By collaborating with environmental advocacy groups, community-based organizations, and government agencies, global healthcare delivery systems can implement:

1. Green hospital infrastructure, including rooftop gardens, sustainable energy sources, and wastewater recycling (HCWH, 2021).
2. Empirically-supported, ecosystem-based disease prevention, such as mangrove restoration for coastal protection against waterborne diseases or urban tree planting to reduce air pollution-related respiratory illnesses (WHO, 2021).
3. NbS-informed medical education at all levels, ensuring current and future healthcare professionals understand the intersection between environmental health protection, disease prevention, health promotion, disease surveillance and population health management, and emergency preparedness, resilience, and response (Whitmee et al., 2015).

These efforts require all healthcare institutions to partner with environmental experts to develop evidence-based NbS strategies that align with traditional clinical medicine best practices, essential public health functions, community-based organization services, and sustainability goals.

7.3.4 Environmental Organizations as Knowledge and Innovation Hubs.

Environmental organizations and advocacy groups provide essential subject matter interdisciplinary expertise in ecological restoration, Anthropocene climate change mitigation and adaptation, and biodiversity conservation, which are foundational to nature-based healthcare services delivery and infrastructure design. Non-governmental organizations (NGOs), academic healthcare institutions, and multi-level environmental organizations and advocacy groups contribute by:

1. Conducting ecological health impact assessments, helping governments and healthcare providers understand how NbS can mitigate disease and improve health outcomes (Cohen-Shacham et al., 2016).

2. Developing innovative NbS interventions, such as integrating wetland restoration with water purification strategies to reduce cholera outbreaks (Kabisch et al., 2017).
3. Providing technical guidance with foundational and capacity-building, ensure that healthcare providers and policymakers can implement NbS effectively (UNEP, 2021).

Collaboration between healthcare and environmental sectors bridges the knowledge gap between biodiversity conservation and traditional clinical medicine, fostering holistic, comprehensive, and integrated approaches to sustainable healthcare service delivery.

7.3.5 Case Studies Highlighting Successful NbS Collaborations.

Several global initiatives demonstrate the impact of interdisciplinary and multisectoral collaboration in NbS adoption:

1. *The Green Climate Fund (GCF) and WHO Climate-Resilient Health Systems Initiative*: Supports climate-adaptive healthcare infrastructure, integrating NbS into hospital designs and national health policies (WHO, 2022).
2. *The Global Green and Healthy Hospitals (GGHH) Network*: A collaboration between hospitals and environmental organizations to reduce GHG emissions, adopt sustainable medical supply chain procurement, and integrate NbS into patient-care settings (HCWH, 2021).
3. *China's Sponge Cities Initiative*: A government-led effort in collaboration with urban planners and environmental scientists to improve stormwater management, reduce flooding risks, and enhance urban health resilience (Li et al., 2021).

These examples highlight the necessity of interdisciplinary and multi-sectoral public private partnerships (PPPs) in successfully implementing NbS-driven healthcare service delivery.

7.3.6 Overcoming Barriers to Effective Collaboration.

Despite the benefits of interdisciplinary and multisectoral partnerships, several challenges hinder effective collaboration in adopting NbS:

1. Fragmented policy frameworks, where environmental and healthy public policies operate in silos rather than integrating NbS as a shared solution (WHO, 2021).
2. Lack of sustainable funding and investment, as many governments and healthcare institutions prioritize short-term medical solutions over long-term ecological resilience (World Bank, 2022).
3. Limited awareness and training, with healthcare professionals and policymakers often lacking expertise in NbS implementation (UNEP, 2021).

To overcome these barriers, governments at all levels of organization, academic and non-academic healthcare institutions, and environmental organizations and advocacy groups must:

1. Develop integrated global and national healthcare strategies that align traditional clinical medicine, essential public health functions, ecosystems' services, and climate-change action.
2. Secure dedicated and sustainable funding for NbS initiatives, leveraging international climate finance mechanisms such as the Green Climate Fund (GCF).
3. Strengthen knowledge-sharing platforms, ensuring interdisciplinary and multisectoral education and research collaborations between public health experts and environmental scientists.

7.3.7 Conclusion.

Interdisciplinary and multisectoral collaboration between governments at all levels of organization, academic and non-academic healthcare institutions, and environmental organizations and advocacy groups is essential for effectively adopting Nature-based Solutions (NbS) in health and environmental sustainability efforts. By aligning both environmental and healthy public policies, integrating ecological subject matter expertise into healthcare planning, and fostering interdisciplinary and multi-sectoral partnerships, these stakeholders can enhance climate-change resilience, reduce chronic disease burdens, and promote sustainable positive health outcomes. Addressing systemic barriers through public policy coordination, investment in NbS programs, and foundational and capacity-building efforts will be crucial in scaling up nature-based healthcare interventions worldwide.

7.4 References.

1. Chivian, E., & Bernstein, A. (2008). *Sustaining life: How human health depends on biodiversity.* Oxford University Press.

2. Cohen-Shacham, E., Walters, G., Janzen, C., & Maginnis, S. (2016). *Nature-based solutions to address global societal challenges.* IUCN.

3. European Commission. (2021). *EU Green Deal and Biodiversity Strategy: Integrating nature-based solutions into health policies.*

4. European Commission. (2021). *The European Green Deal and health implications.*

5. Health Care Without Harm (HCWH). (2021). *Global Green and Healthy Hospitals Network: Promoting sustainable healthcare systems.*

6. Health Care Without Harm (HCWH). (2021). *Global roadmap for healthcare decarbonization.*

7. Kabisch, N., Korn, H., Stadler, J., & Bonn, A. (Eds.). (2017). *Nature-based solutions to climate change adaptation in urban areas.* Springer.

8. Kickbusch, I., Gleicher, D., & Pozo-Martin, F. (2019). Global health diplomacy: A new era of leadership and governance. *BMJ Global Health, 4*(Suppl 3), e001883.

9. Li, Y., Borthwick, A. G. L., & Bao, H. (2021). Sponge city concept helps solve China's urban flooding. *Nature Sustainability, 4*(8), 665-667.

10. Romanello, M., McGushin, A., Di Napoli, C., et al. (2021). The 2021 report of the Lancet Countdown on health and climate change: Code red for a healthy future. *The Lancet, 398*(10311), 1619-1662.

11. The Lancet. (2021). The Glasgow Climate Pact: A fragile win for health. *The Lancet, 398*(10316), 2061.

12. Ulrich, R. S., Simons, R. F., Losito, B. D., Fiorito, E., Miles, M. A., & Zelson, M. (2008). Stress recovery during exposure to natural and urban environments. *Journal of Environmental Psychology, 11*(3), 201-230.

13. United Nations Development Programme (UNDP). (2019). *Climate action and the SDGs: Synergies and trade-offs.* United Nations Development Programme.

14. United Nations Environment Programme (UNEP). (2020). *Integrating nature-based solutions into sustainable development goals (SDGs).* United Nations Environment Programme.

15. United Nations Environment Programme (UNEP). (2020). *Sustainable Development Goals and the environment.* United Nations Environment Programme.

16. United Nations Environment Programme (UNEP). (2021). *Making peace with nature: A scientific blueprint to tackle the climate, biodiversity, and pollution emergencies.* United Nations Environment Programme.

17. United Nations Framework Convention on Climate Change (UNFCCC). (2015). *The Paris Agreement.* United Nations Framework Convention on Climate Change.

18. United Nations Framework Convention on Climate Change (UNFCCC). (2021). *Glasgow Climate Pact: COP26 outcomes.*

19. Watts, N., Amann, M., Arnell, N., et al. (2021). The 2021 report of the Lancet Countdown on health and climate change. *The Lancet, 398*(10311), 1619-1662.

20. Whitmee, S., Haines, A., Beyrer, C., et al. (2015). Safeguarding human health in the Anthropocene epoch: Report of The

Rockefeller Foundation–Lancet Commission on planetary health. *The Lancet, 386*(10007), 1973-2028.

21. World Health Organization (WHO). (2016). *Global strategy on human resources for health: Workforce 2030.*

22. World Health Organization (WHO). (2020). *WHO manifesto for a healthy recovery from COVID-19.* World Health Organization.

23. World Health Organization (WHO). (2021). *WHO operational framework for building climate-resilient health systems.* World Health Organization.

24. World Health Organization (WHO). (2022). *Nature-based solutions and health: Policy recommendations for integrating ecosystem services into healthcare.* World Health Organization.

25. World Health Organization (WHO). (2022). *Health in the Nationally Determined Contributions (NDCs): An analysis of how climate commitments address health.* World Health Organization.

26. World Bank. (2022). *Investing in nature for development and climate resilience.*

27. World Bank. (2022). *Investing in climate-resilient health systems: A priority for global health security.* World Bank Group.

8.0

Conclusion.

8.1 The Interconnectedness of Anthropocene Climate Change, Global Healthcare Service Delivery, EHS, and Sustainability.

8.1.1 Introduction.

The interconnectedness of Anthropocene climate change, healthcare service delivery, Environmental Health and Safety (EHS), and Sustainability—encompassing Environmental, Social, and Governance (ESG) principles—underscores the growing recognition that environmental sustainability is fundamental to the public's health, wellness, and well-being, corporate social responsibility, and global resilience. Climate change intensifies health risks, contributing to chronic diseases and increasing the demand for robust healthcare infrastructure, such as hospitals and emergency departments. It also necessitates enhanced workplace safety measures and stronger corporate governance to support sustainable environmental and public health policies. Integrating these domains enables healthcare institutions, businesses, and governments to align environmental protection with traditional medical care, essential public health services, and ethical governance standards (Watts et al., 2021).

8.1.2 Climate Change and Global Healthcare Delivery: A Direct Link to Public Health Crises.

Human-induced climate change has profound consequences for human health, with rising global temperatures, extreme weather events, and environmental degradation contributing to an increase in:

1. Vector-borne diseases, such as malaria and dengue fever, due to changing habitats for disease-carrying insects (WHO, 2021).
2. Respiratory illnesses, exacerbated by worsening air pollution and wildfires (Romanello et al., 2022).
3. Waterborne diseases, linked to flooding and contaminated water sources (UNEP, 2021).
4. Mental health challenges, driven by climate-related displacement, food insecurity, and disasters (Gifford & Gifford, 2016).

Global healthcare delivery systems must adapt to climate-related health burdens by implementing resilient biophilic infrastructure design and implementation, sustainable medical and public health practices, and disaster/pandemic preparedness programs (WHO, 2021).

8.1.3 EHS: Protecting Workers and Communities Amid Climate Change.

Environmental Health and Safety (EHS) programs focus on workplace safety, environmental protection, and public health initiatives, making them integral to climate change adaptation and mitigation. Climate-related hazards impact:

1. Workplace health, with increased risks of heat stress, air pollution exposure, and disaster-related injuries for frontline workers (ILO, 2019).

2. Industrial sustainability, where businesses must mitigate pollution, waste, and hazardous materials affecting both employee and community health (OSHA, 2021).
3. Public health resilience, ensuring that workplace safety measures align with broader community health goals in the face of environmental challenges (US EPA, 2022).

EHS strategies must integrate climate risk assessments, sustainable resource management, and pollution control to protect both workers and communities.

8.1.4 Sustainability: Corporate Responsibility for Health and Climate Action.

Sustainability frameworks guide organizations in adopting sustainable and ethical practices, influencing business strategies that impact both the environment and the public's health. Foundational sustainability considerations related to healthcare and climate change include:

1. *GHG reductions*: Companies are expected to decrease GHG emissions from operations, supply chains, and product lifecycles with a goal of becoming net-zero by 2030 (SBTi, 2021).
2. *Social responsibility in healthcare and non-healthcare organizations*: Companies investing in employee wellness and well-being, essential health services and benefits, and behavioral (i.e., substance use disorder and mental health services) health support align with sustainability's corporate social responsibility pillar (OECD, 2022).
3. *Governance for sustainability*: Transparent, truthful, and trustworthy reporting, multi-level regulatory compliance, and climate-related risk disclosures ensure ethical environmental and healthcare practices (TCFD, 2021).

Global initiatives, such as the Task Force on Climate-Related Financial Disclosures (TCFD) and Sustainable Development Goals (SDGs), drive corporations to align key sustainability measures with climate and healthcare-related commitments (UNEP FI, 2022).

8.1.5 The Need for Integrated Healthy Public Policies and Multisectoral Collaboration.

The synergistic relationship between Anthropocene climate change, healthcare service delivery, Environmental Health and Safety (EHS), and Sustainability—encompassing Environmental, Social, and Governance (ESG) principles—underscores the necessity of creating integrated healthy public policies that promote sustainability principles, positive healthcare outcomes, and corporate social responsibility. Strategies to enhance this interconnected and multisectoral approach include:

1. Developing climate-resilient global healthcare delivery systems, incorporating green infrastructure, renewable energy, and circular economy principles (WHO, 2021).
2. Strengthening compliance and regulatory frameworks, ensuring businesses adhere to EHS and Sustainability standards that mitigate environmental and health risks leading to evidence-based risk management (EBRM) (US EPA, 2022).
3. Encouraging public-private partnerships (PPPs), fostering collaboration between governments, NGOs, academic and non-academic healthcare institutions, and corporations for climate-health initiatives (HCWH, 2021).
4. Promoting sustainability reporting, requiring organizations to disclose their environmental impact and health-related risk mitigation strategies (TCFD, 2021).

8.1.6 Conclusion.

The interconnectedness of Anthropocene climate change, healthcare service delivery, Environmental Health and Safety (EHS), and Sustainability—including Environmental, Social, and Governance (ESG) principles—underscores the urgent need for a holistic, interdisciplinary, and multisectoral approach to sustainability and health resilience. Climate change exacerbates health risks by increasing the prevalence of chronic diseases, vector-borne illnesses, respiratory conditions, and heat-related morbidity, necessitating stronger EHS protections and sustainability-driven corporate accountability to safeguard both workers and communities.

A proactive approach to integrating climate resilience into global healthcare delivery systems requires investment in sustainable infrastructure design and implementation, climate-adaptive healthcare and environmental public policies, and environmentally responsible medical supply chain management. Strengthening compliance and regulatory frameworks, fostering cross-sector collaboration, and prioritizing sustainable innovation are essential to mitigating environmental degradation while ensuring equitable access for all to healthcare services delivery.

By aligning environmental policies with the long-term sustainability of global healthcare delivery systems and ethical corporate governance, stakeholders—including governments, healthcare institutions, businesses, and civil society—can advance a climate-smart, health-conscious future. This alignment not only protects the public's health but also enhances corporate responsibility, economic stability, and ecosystem services preservation, ultimately benefiting people, businesses, and the planet.

8.2 The Potential for NbS to Revolutionize the Healthcare Sector and the Medical-Industrial Complex.

8.2.1 Introduction.

The global healthcare sector and the broader medical-industrial complex face growing sustainability challenges, including high GHG emissions, resource overconsumption, and environmental degradation (Karliner et al., 2020). Nature-based Solutions (NbS) offer a transformational approach to improving healthcare infrastructure design, healthcare outcomes, and system-wide sustainability by integrating ecosystems' services into healthcare services delivery and operations. By leveraging ecosystem processes, NbS can help mitigate climate change, enhance the public's health, and promote resource efficiency in the global medical and healthcare industry (WHO, 2022).

8.2.2 Reducing the Environmental Impact of Healthcare Infrastructure.

The healthcare sector is responsible for 4.4 % of global greenhouse gas (GHG) emissions, primarily from energy-intensive hospital operations, medical supply chains, and waste management (Health Care Without Harm, 2019). NbS can reduce this footprint by:

1. Incorporating green infrastructure, such as rooftop gardens, urban forests, and biophilic hospital designs, which improve air quality, regulate temperature, and reduce energy consumption (Houghton & Castillo-Salgado, 2021).
2. Utilizing sustainable water management systems, including constructed wetlands and rainwater harvesting, to decrease hospital water consumption and pollution (UNEP, 2021).

3. Designing nature-integrated healing environments, which reduce hospital stays and improve patient recovery by enhancing psychological well-being (Ulrich et al., 2020).

8.2.3 Enhancing Public and Proactive Health Through NbS.

NbS support proactive healthcare by addressing environmental determinants of health. Climate change, pollution, and biodiversity loss exacerbate chronic diseases, infectious outbreaks, and mental health disorders. Integrating NbS can help:

1. Combat air pollution-related illnesses by increasing urban green spaces that reduce particulate matter and respiratory disease incidence (Nowak et al., 2018).
2. Prevent heat-related health issues through nature-based cooling solutions, such as tree canopies and urban wetlands, which lower urban heat island effects (Oke et al., 2017).
3. Improve mental health and well-being by expanding access to green spaces, promoting ecotherapy, and integrating nature into healthcare interventions (Bratman et al., 2019).

8.2.4 Sustainable Global Medical Supply Chains and Biopharmaceutical Innovations.

The global medical-industrial complex relies heavily on resource-intensive supply chains. NbS can revolutionize manufacturing, product life-cycle, and waste management by:

1. Advancing sustainable pharmaceuticals, where plant-based and microbial solutions offer biodegradable, less toxic alternatives to synthetic drugs (Newman & Cragg, 2020).
2. Promoting circular economy models, such as biodegradable medical materials and nature-inspired packaging solutions that reduce hospital waste (World Bank, 2020).

3. Integrating bioengineered solutions, including nature-derived biomaterials for wound care, regenerative medicine, and biodegradable implants (Chakraborty et al., 2022).

8.2.5 Climate Resilience and Health Equity through NbS.

Vulnerable communities suffer disproportionately from climate-related health crises, including flooding, food insecurity, and vector-borne diseases. NbS can enhance global healthcare delivery system resilience by:

1. Designing climate-adaptive healthcare facilities, such as hospitals with flood-resistant green infrastructure and energy-efficient designs (WHO, 2021).
2. Supporting nature-based public health initiatives, including agroforestry and community gardens that enhance food security and nutrition (FAO, 2019).
3. Strengthening ecosystem-based disease prevention, such as mangrove and wetland conservation to reduce malaria and waterborne disease risks (UNDP, 2022).

8.2.6 Policy and Governance: Scaling NbS in Healthcare.

Despite their benefits, NbS remain underutilized in global healthcare service delivery due to public policy gaps and financial barriers. Governments, global academic and non-academic healthcare institutions, and international environmentally-focused organizations and advocacy groups must:

1. Incorporate NbS into national healthy public policies, aligning sustainability principles with climate adaptation goals (UNEP, 2022).
2. Invest in research and innovation, funding nature-based clinical medical solutions and green hospital designs (Health Care Without Harm, 2021).

3. Develop multi-sector partnerships, engaging healthcare institutions, urban planners, and environmental organizations in NbS implementation (WHO & UNDP, 2021).

8.2.7 Conclusion.

Nature-Based Solutions (NbS) offer a groundbreaking opportunity to transform the global healthcare sector and the medical-industrial complex by integrating environmental sustainability with healthcare system and service delivery innovation. From reducing healthcare emissions and enhancing environmentally-friendly patient-centered care to creating climate-resilient healthcare delivery systems worldwide, NbS present a holistic, cost-effective, and ecologically responsible path forward. To fully harness their potential, health leaders, policymakers, and industry stakeholders must prioritize NbS adoption, fostering a sustainable and healthcare-centered resilient environment for today and tomorrow.

8.3 Encouragement for Global Healthcare Delivery Systems to Lead the Charge in Adopting NbS for a Healthy Planet.

8.3.1 Introduction.

As the global healthcare sector continues to confront the growing threats of Anthropocene climate change, environmental degradation, and public health crises (e.g., COVID-19 pandemic), global healthcare delivery systems must take a leadership role in adopting Nature-Based Solutions (NbS) to promote planetary and human health. All healthcare systems have a dual responsibility—not only to treat human illnesses but also to mitigate the environmental factors exacerbating disease burdens. By integrating NbS, the global healthcare community can drive a paradigm shift toward a sustainable, resilient, and health-centered future (WHO, 2022).

8.3.2 The Urgency for Healthcare Leadership in Sustainability.

As previously noted, the healthcare sector contributes nearly 4.4% of global GHG emissions, disproportionately impacting marginalized and vulnerable populations suffering from air pollution, climate-related illnesses, and ecosystem degradation (Health Care Without Harm, 2019). The sector's reliance on energy-intensive operations, excessive waste production, and resource overconsumption necessitates a fundamental rethinking of healthcare infrastructure design and services delivery (Karliner et al., 2020). Healthcare leaders must advocate for NbS integration to counteract these challenges by promoting:

1. Green hospital infrastructure, including green roofs, tree canopies, and energy-efficient designs to reduce environmental impact (Houghton & Castillo-Salgado, 2021).
2. Sustainable water management, such as rainwater harvesting and constructed wetlands, to enhance hospital resilience to water shortages and contamination risks (UNEP, 2021).
3. Eco-friendly medical supply chains, incorporating biodegradable materials and nature-derived pharmaceuticals to minimize toxic waste and environmental harm (Newman & Cragg, 2020).

8.3.3 NbS as a Public Health Strategy.

Global healthcare delivery systems must go beyond traditional clinical medicine treating acute and chronic health conditions and address the root causes of health disparities linked to adverse environmental impact. NbS offer scalable interventions to reduce healthcare burdens by:

1. Reducing urban air pollution, which is responsible for millions of premature deaths annually, by increasing urban tree cover and green spaces (Nowak et al., 2018).

2. Enhancing mental health outcomes through exposure to nature, which has been linked to reduced anxiety, depression, and stress-related disorders (Bratman et al., 2019).
3. Combatting heat-related illnesses by deploying urban forests and wetland restoration projects that cool urban environments and prevent heatstroke (Oke et al., 2017).

8.3.4 Policy and Institutional Support for NbS Adoption.

Governments, NGOs, environmentally-focused organizations and advocacy groups, and both academic and non-academic healthcare institutions must align their sustainability commitments with global climate and healthy public policies, such as the Paris Agreement (2015) and the WHO Operational Framework for Climate-Resilient Health Systems (2021). Key policy actions include:

1. Embedding NbS into national healthcare delivery strategies to align healthcare sustainability goals with climate resilience frameworks (UNDP, 2022).
2. Incentivizing hospitals and clinics to implement green infrastructure through subsidies, tax incentives, and sustainability certification programs (World Bank, 2020).
3. Encouraging interdisciplinary and multisectoral collaboration between global healthcare providers, international environmental agencies, and urban planners to scale NbS adoption (WHO & UNDP, 2021).

8.3.5 Scaling Global Healthcare's Role in Climate Resilience and Health Equity.

As frontline responders to climate-related disasters and pandemics, healthcare delivery systems must build climate resilience while ensuring equitable access to NbS benefits, especially in underserved communities. This can be achieved by:

1. Developing climate-adaptive healthcare facilities that integrate NbS, such as flood-resistant hospitals and renewable energy-powered clinics (WHO, 2021).
2. Supporting nature-based food security initiatives, including hospital-run community gardens and agroforestry programs that enhance nutrition and local food systems (FAO, 2019).
3. Integrating ecosystem-based disease prevention by restoring wetlands and biodiversity conservation to reduce the spread of vector-borne diseases like malaria and dengue (UNEP, 2021).

8.3.6 Call to Action: Healthcare as a Global Leader in Sustainability.

The global healthcare sector has both the moral and strategic imperative to lead the world-wide movement toward a healthier planet and population. By adopting NbS, global healthcare institutions can:

1. Set a precedent for other industries, demonstrating the feasibility and benefits of nature-based interventions.
2. Influence global climate and healthy public policy, advocating for sustainable funding and investments into integrated systems of health with increased spending for primary health care.
3. Empower communities through education, outreach, and implementation of nature-driven health solutions.

8.3.7 Conclusion.

Global healthcare delivery systems must lead the charge in adopting Nature-based Solutions (NbS) as a core strategy for planetary and public health. By reducing environmental harm, enhancing resilience, and promoting health equity, the global healthcare sector can drive a major paradigm shift toward sustainable, climate-smart healthcare services delivery world-wide. The time for action is now—all global healthcare institutions must become catalysts for NbS integration, ensuring a healthier world for both today and for future generations to come.

8.4 Inspiring Healthcare Stakeholders to Advocate for and Implement NbS in Local Healthcare Delivery Systems.

8.4.1 Introduction.

As Anthropocene climate change and environmental degradation continue to threaten the public's health, healthcare stakeholders— including policymakers, hospital administrators, clinicians, research- ers, and community health advocates—must embrace Nature-based Solutions (NbS) as an essential component of sustainable healthcare service delivery. By integrating NbS, all academic and non-academic healthcare institutions can mitigate environmental harm, enhance patient outcomes, and strengthen system resilience, while leading the global charge toward climate-smart, nature-driven integrated person-centered systems of health (WHO, 2022).

8.4.2 The Moral and Ethical Imperative for NbS in Healthcare.

Healthcare professionals hold a fundamental responsibility to "do no harm", which extends beyond patient care to include environmental stewardship (Karliner et al., 2020). The healthcare sector is a major contributor to pollution, accounting for nearly 4.4 % of global GHG emissions, excessive resource consumption, and high levels of waste production (Health Care Without Harm, 2019).

1. Adopting NbS, such as green hospital infrastructure design and sustainable medical supply chains, aligns with ethical commit- ments to protect the public health and environmental integrity (Newman & Cragg, 2020).
2. Reducing air and water pollution through urban greening and wetland restoration minimizes disease burdens linked to respira- tory illnesses and waterborne infections (WHO, 2021).

8.4.3 Empowering Healthcare Leaders and Decision-Makers.

Hospital administrators, health policymakers, and institutional leaders must spearhead the integration of NbS into healthcare infrastructure design, governance, and service delivery (World Bank, 2020). Key initiatives include:

1. Green hospital design, incorporating rooftop gardens, energy-efficient buildings, and natural ventilation to reduce heat stress, air pollution, and GHG emissions (Houghton & Castillo-Salgado, 2021).
2. Sustainable procurement policies, prioritizing biodegradable medical supplies, nature-derived pharmaceuticals, and eco-friendly durable medical equipment to reduce environmental toxicity (UNEP, 2021).
3. Decentralized renewable energy solutions, such as solar-powered healthcare facilities and nature-driven cooling systems, to increase climate resilience and operational sustainability (UNDP, 2022).

8.4.4 Clinicians as Champions of NbS-Driven Healthcare.

Healthcare providers—including physicians, nurses, and allied health professionals—play a pivotal role in advocating for environmentally conscious medical practices that prioritize both human and planetary health (WHO & UNDP, 2021). Clinicians can:

1. Promote nature-based therapies, such as ecotherapy and green prescriptions, which have been shown to improve mental health, cardiovascular health, and chronic disease management (Bratman et al., 2019).
2. Educate patients on environmental determinants of health, including the health benefits of exposure to green spaces, clean air, and sustainable diets (FAO, 2019).

3. Implement waste reduction strategies, such as sustainable anesthesia practices and single-use plastic reduction in operating rooms, to minimize the ecological footprint of healthcare services (Karliner et al., 2020).

8.4.5 Engaging Communities in NbS-Driven Healthcare Interventions.

Local communities must be empowered to take an active role in implementing NbS-driven healthcare interventions, ensuring equitable access to green infrastructure and nature-based health benefits (WHO, 2022). Community-driven initiatives include:

1. Urban reforestation projects to reduce urban heat islands and improve respiratory health outcomes (Nowak et al., 2018).
2. Nature-based mental health programs, such as community gardens and forest therapy, to combat stress, depression, and anxiety (Bratman et al., 2019).
3. Integrated health and agriculture programs, supporting agroecological food systems that improve nutrition and reduce diet-related diseases (FAO, 2019).

8.4.6 Healthy Public Policy Advocacy and Global Collaboration.

Governments and international healthcare organizations must institutionalize NbS within healthy public policies and sustainable funding mechanisms to drive large-scale transformation (UNDP, 2022). Stakeholders can:

1. Advocate for national and international policy alignment with frameworks such as the Paris Agreement (2015) and the WHO Operational Framework for Climate-Resilient Health Systems (2021) (WHO, 2021).

2. Foster cross-sector collaborations, linking healthcare institutions with environmental agencies, urban planners, and civil society organizations to co-design NbS strategies (World Bank, 2020).
3. Secure investment and financial incentives, such as sustainability grants, tax credits, and green bonds, to fund sustainable NbS-driven healthcare infrastructure (UNEP, 2021).

8.4.7 Call to Action: A Unified Movement Toward Sustainable Global Healthcare Delivery Systems.

For NbS to become a global standard in global healthcare delivery systems, stakeholders at all levels must unite to drive change. The global healthcare sector must:

1. Champion a proactive shift toward sustainability by embedding NbS into healthcare facility operations, traditional clinical medicine, and essential public health functions.
2. Lead by example, demonstrating how NbS can improve patient-centered care while reducing environmental impact.
3. Engage policymakers, communities, and industries in a collective effort to transform global healthcare services delivery into a healthcare model for environmental and human health synergy.

8.4.8 Conclusion.

The global healthcare sector stands at a pivotal moment, with an unprecedented opportunity to redefine its relationship with nature by integrating Nature-based Solutions (NbS) as a foundational principle of sustainable healthcare. In an era marked by Anthropocene climate change, biodiversity loss, and increasing environmental health risks, healthcare service delivery must move beyond traditional models of care delivery and actively embrace environmental stewardship.

To achieve this transformation, all stakeholders—including healthcare leaders, clinicians, policymakers, researchers, and communities—must advocate for and implement NbS at every level of healthcare delivery. This means designing hospitals and clinics that incorporate green infrastructure, promoting nature-based therapies to enhance patient well-being, and adopting sustainable resource management practices to minimize the sector's environmental footprint. Furthermore, integrating NbS into public health and healthy public policies can strengthen climate resilience, mitigate the spread of zoonotic diseases, and reduce healthcare disparities, particularly among vulnerable and marginalized populations disproportionately affected by environmental degradation.

By embedding NbS into global healthcare delivery systems, the sector will not only improve patient outcomes through holistic, proactive, and restorative approaches but also contribute to the long-term health of the planet. This commitment to ecological and human health interconnectivity is essential for ensuring that future generations inherit a world where healthcare thrives in harmony with nature.

8.5 References.

1. Bratman, G. N., Hamilton, J. P., Hahn, K. S., Daily, G. C., & Gross, J. J. (2019). Nature and mental health: An ecosystem service perspective. *Science Advances, 5*(7), eaax0903.
2. Chakraborty, S., Das, T., & Ghosh, S. K. (2022). Biodegradable materials for medical applications: A sustainable approach. *Journal of Applied Biomaterials & Functional Materials, 20*(1), 1-15.
3. Food and Agriculture Organization (FAO). (2019). *The state of the world's biodiversity for food and agriculture.* FAO.
4. Gifford, R., & Gifford, S. (2016). The largely unrecognized impact of climate change on mental health. *Bulletin of the Atomic Scientists, 72*(5), 292-297.
5. Health Care Without Harm (HCWH). (2019). *Health care's climate footprint: How the health sector contributes to the global climate crisis and opportunities for action.*

6. Health Care Without Harm (HCWH). (2021). *Global roadmap for healthcare decarbonization.*

7. Health Care Without Harm (HCWH). (2021). *A roadmap to sustainable healthcare.*

8. Houghton, A., & Castillo-Salgado, C. (2021). Green hospital design and its impact on patient well-being. *Journal of Environmental Health, 84*(2), 10-17.

9. International Labour Organization (ILO). (2019). *Working on a warmer planet: The impact of heat stress on labour productivity and decent work.* ILO.

10. Karliner, J., Slotterback, S., Boyd, R., Ashby, B., & Steele, K. (2020). *Climate action in healthcare: Measuring and reducing the environmental impact of healthcare operations.* Health Care Without Harm.

11. Newman, D. J., & Cragg, G. M. (2020). Natural products as sources of new drugs over the nearly four decades from 1981 to 2019. *Journal of Natural Products, 83*(3), 770-803.

12. Nowak, D. J., Hirabayashi, S., Bodine, A., & Greenfield, E. (2018). Tree and forest effects on air quality and human health in the United States. *Environmental Pollution, 242*(A), 90-101.

13. Oke, T. R., Mills, G., Christen, A., & Voogt, J. A. (2017). *Urban climates.* Cambridge University Press.

14. Organisation for Economic Co-operation and Development (OECD). (2022). *ESG investing and climate risk disclosure.*

15. Occupational Safety and Health Administration (OSHA). (2021). *Climate change and worker safety.*

16. Romanello, M., McGushin, A., Di Napoli, C., et al. (2022). The 2022 report of the Lancet Countdown on health and climate change. *The Lancet, 400*(10363), 1619-1662.

17. Science-Based Targets Initiative (SBTi). (2021). *Corporate net-zero standard: Science-based pathways for businesses.*

18. Task Force on Climate-Related Financial Disclosures (TCFD). (2021). *Guidance on climate-related financial disclosures.*

19. Ulrich, R. S., Zimring, C. M., Zhu, X., DuBose, J., Seo, H. B., Choi, Y. S., ... & Joseph, A. (2020). A review of the impact of green spaces in healthcare settings. *HERD: Health Environments Research & Design Journal, 13*(3), 14-29.

20. United Nations Development Programme (UNDP). (2022). *Nature-based solutions for health resilience.*

21. United Nations Environment Programme (UNEP). (2021). *Making peace with nature: A scientific blueprint to tackle the climate, biodiversity, and pollution emergencies.* United Nations Environment Programme.

22. United Nations Environment Programme (UNEP). (2021). *Nature-based solutions for sustainable development.*

23. United Nations Environment Programme Finance Initiative (UNEP FI). (2022). *The role of finance in the climate-health nexus.* Retrieved from

24. United States Environmental Protection Agency (US EPA). (2022). *Climate change and public health.*

25. Watts, N., Amann, M., Arnell, N., et al. (2021). The 2021 report of the Lancet Countdown on health and climate change. *The Lancet, 398*(10311), 1619-1662.

26. World Bank. (2020). *Innovations in circular economy for healthcare waste management.*

27. World Health Organization (WHO). (2021). *WHO operational framework for building climate-resilient health systems.*

28. World Health Organization (WHO). (2021). *Building climate-resilient health systems.*

29. World Health Organization (WHO). (2022). *Nature-based solutions and human health: The role of ecosystems in supporting health and well-being.*